Monitoring Continuous Phenomena

Background, Methods and Solutions

Peter Lorkowski

Formerly Researcher & Lecturer
Institute for Applied Photogrammetry and Geoinformatics (IAPG)
Jade University of Applied Sciences, Oldenburg, Germany

Senior Software Developer
Devity Labs GmbH, Oldenburg, Germany

CRC Press
Taylor & Francis Group
Boca Raton   London   New York

CRC Press is an imprint of the
Taylor & Francis Group, an **informa** business

A SCIENCE PUBLISHERS BOOK

First edition published 2021
by CRC Press
6000 Broken Sound Parkway NW, Suite 300, Boca Raton, FL 33487-2742

and by CRC Press
2 Park Square, Milton Park, Abingdon, Oxon, OX14 4RN

© 2021 Taylor & Francis Group, LLC

CRC Press is an imprint of Taylor & Francis Group, LLC

ISBN: 978-1-138-33973-6 (hbk)
ISBN: 978-0-367-77516-2 (pbk)
ISBN: 978-0-429-44096-0 (ebk)

Typeset in Times New Roman
by Radiant Productions

To my wife Christina
and my daughter Elisa

Foreword

The monitoring of our environment is an aspect of science that is becoming increasingly important. Since mankind has gained the power to exploit the environment in order to achieve goals (which were and are not always to the benefit of all), the impact on it has become so immense that it has the potential to change the complex system *earth* in a way that will negatively affect the conditions under which future generations will have to live.

One important example in history is the harmful effect that chlorofluorocarbon (CFC) have on the ozone layer of our atmosphere. The problem was detected by scientists in 1973, propagated and politically absorbed and responded to by a worldwide agreement through the Montreal Protocol in 1989.[1]

On a timescale of decades, the effects of the CFC ban could be observed. In 2013, the trend toward recovery of the ozone layer could be confirmed. The causal relation between CFC emission and ozone layer destruction has thus been demonstrated in this global "experiment".

The transformation from scientific discovery to political action was successful and had a positive effect on the emissions. The assured knowledge about the effects of CFC to the atmosphere was necessary to trigger the collective endeavour of the world's nations. It finally made life on earth safer.

The reason for this success story was the relative simplicity and traceability of the relation between cause (CFC emissions) and effect (depletion of ozone layer). It was also due to the rather limited economic impact that was caused by the necessary change: it was only a small industrial branch that was affected and technological alternatives could soon be provided.

Nevertheless, there has been significant resistance by the industry against the CFC ban. However, given the constellation of has a high degree of scientific

---

[1] https://ozone.unep.org/ozone-timeline, visited 2020-01-06.

certainty, the severe impact on mankind (higher risk of skin cancer) and limited power of social groups that would benefit from continuing production, it was implemented anyway.

To our knowledge, the most severe environmental problem of our times is climate change caused by human activity. The impact will be much more severe than that of the ozone destruction could ever be. The scientific evidence is almost overwhelming, but climate is far more complex system which makes it quite easy for opportunistic denialists to cast doubt and thus hinder collective progress in reducing greenhouse gas emissions.

The industrial complex with its immense resources accumulated in the last decades is tremendously powerful and the average citizen is overburdened with judging the validity of differing or even contradicting scientific statements about climate change. Denialist claims yield attention that is not at all proportionate to their appreciation within the scientific community.

Given these complex and unfortunate circumstances makes the protection of the environment a much more difficult endeavour than it has been in the past. Immense capital has been accumulated which can now be used to feed a machinery designed to retain the greenhouse gas-emitting cash cows. Short-term profits are still preferred to the necessary sustainable transformation of industry and society.

Yet, there is a rising awareness about man-made climate change. An increasing frequency of extreme events like heat waves, droughts, fires, storms and heavy rainfall make it more difficult to deny it. The cost of a missing regulation on a greenhouse gas emissions becomes more and more apparent and it becomes obvious that it might very well surpass the cost of reducing them by large magnitudes.

Notwithstanding the still noticeable scepticism towards science, environmental monitoring will remain the only means to feed climate models and simulate future scenarios. Such simulations will always carry a certain degree of uncertainty with them and responsible scientists are therefore usually reluctant with respect to their own predictions. They are aware of the fact that there is still only limited knowledge that science has about such a tremendously complex system like climate. Given the *economy of attention* in the media today, such reluctance is of course a disadvantage when opposed to those arguing against climate change with great confidence, notwithstanding their often inferior expertise.

So there is a growing tension between the rather cautious professionals of science, well aware of its limits, and the increasingly vociferous appearance of political actors. Scientific hypotheses—notwithstanding or actually because of their essential property of potentially being refuted—do represent the best available knowledge about existential environmental threats.

A hypothesis (like e.g. man-made climate change) can never be *fully* proven due to the scientific method per se. To therefore treat this hypothesis as false, though, is a highly irrational strategy in the light of the grave consequences it predicts. It is, nonetheless, the strategy that is *in effect* still pursued by the global population.

Giving scientific insights more weight will be crucial to induce effective environmental protection. This is not only a scientific and/or political endeavour, but to a significant degree also a cultural one.

The central challenge for science in this context is to develop sophisticated models of the crucial aspects of our environment in order to make sound predictions.

The concept of a continuous field is a very powerful model to represent environmental phenomena like temperature, air quality, water salinity or soil properties. It is based on a limited number of discrete observations, but provides estimations of the observed variable at arbitrary positions.

The quality of the representational model depends on that of the monitoring process and the quantity of resources dedicated to it. As is the case for many aspects, the art of a good compromise should be the driving force here.

Throughout the book, the evaluation and comparison of such compromises by several indicators is the leading principle. The introduced framework only covers a small portion of the very complex subject of spatio-temporal interpolation. There is plenty of other literature dealing with this subject. This work rather attempts to grasp the problem of monitoring continuous phenomena as a whole and provides tools designed to address some challenges in this context.

The author was reluctant to rely on existing geostatistical products like gstat written in the programming language R. Instead, he preferred to use the fullfledged language C# in favour of flexibility, portability and full architectural control. The intention behind this choice is a strong cohesion between the abstract concept of the framework and the implementation. The crucial mechanism for evolutionary development of the framework is the relatedness between methods and their parameters as input, and quality indicators as output. The specific features necessary in this context are the following:

- mechanism for generating diverse stationary random fields as reference models
- strong relatedness between reference models and interpolation results
- formal terminology for the assignment of methods and parameters
- a mechanism for automatic systematic variation of methods and parameters
- quantification of computational effort
- abstract definition of complex states of continuous phenomena
- quantification of achieved interpolation quality

Given these features, the framework is capable of applying and evaluating different methods and parameters with respect to the relation between the assigned resources and the achieved model quality. Beyond the mere application of specific methods, it provides the concepts and tools to systematically support the search for the most appropriate method for a specific continuous phenomenon. It does so by striving for a systematic experimental arrangement with an associated set of meaningful and generic target criteria.

Preface

This book is a result of Thomas Brinkhoff asking me for my participation in a research project dealing with complex event processing associated with sensor data. The intended conclusion of the project was a dissertation, which was funded by the *Jade2Pro* program of the Jade University in Oldenburg, Germany. Although the subject is in the domain of data stream management, I was allowed to study it from a geostatistics context. Knowing the method from a course I taught for a couple of years, I found it particularly useful for addressing problems of monitoring continuous phenomena with local sensors. Maybe the most seminal idea for the project was to exploit the outstanding feature of the geostatistical method of kriging: the kriging variance. It is unique among interpolation methods and is a valuable indicator that can be used to tackle various problems of monitoring. The approaches that were developed based on this idea needed to be evaluated, which was the initial inducement for developing an extensive simulation framework. As the project progresses, this framework has evolved as a generic tool for systematic variation and evaluation of methods and parameters to find optimal configurations guided by key indicators.

Besides this technical evolution, a more generic perspective on the problem of monitoring became apparent. This perception coincided with the proposal from CRC Press for this book. At the time I was confident that the material produced so far was a good basis for a book, but I was also aware of significant changes and supplements that were necessary in order to address a broader readership. Fortunately, the *perspective* part of my dissertation already contained hints of what was missing to cover the subject appropriately for the book: an abstract formalism to describe a particular state of a continuous field and a reliable monitoring methodology to continuously determine whether such a state is prevailing or not. When extending the simulation framework in order to carry out such advanced scenarios experimentally, there was a recognition for the need to specifically address vector-raster-interoperability in the context of the monitoring of continuous phenomena.

Covering this subject in writing significantly changed my perspective towards a more abstract mental model of the problem of monitoring. Together with the proposed architecture and the approaches addressing specific problems, I hope that readers will find in this book is a useful orientation amidst the complexity of the covered topic. I am very grateful for the opportunity to share my thoughts about it and would be pleased if it can induce the kind of enjoyment I experienced when my understanding of an abstract concept was significantly improved by a book.

Acknowledgements

For the greatest part, this book is based on my dissertation. In order to provide benefits to a broader readership, some ideas that could only be sketched there needed to be developed further significantly. These ideas were aimed at a generic framework to express conditions and states in the context of continuous phenomena. For an experimental evaluation of the concepts, the simulation framework as developed so far needed to be extended by various features.

From the beginning of the book project, I was convinced that these extensions were crucial to produce a more generic monograph from a rather specific dissertation. However, as is almost always the case for software development, I strongly underestimated the effort that was necessary to actually implement the missing features. Various discussions about the benefits of the approaches were very helpful to keep up the motivation to finally implement, test, and evaluate the concepts at hand. This was one building block for the holistic coverage of the theme as was attempted by this work.

In the course of my work about monitoring of continuous phenomena I was inspired by many other authors off which Noel Cressie stands out with his rather early work *Statistics for spatial data*. Without even roughly grasping the complexity of this subject matter at that time, this book somehow sparked an excitement about the potential of this approach which was corroborated in the course of my work.

On a personal level, there have been various people I actually met and want to mention here because the interaction with them influenced my work in some significant way.

My first confrontation with geostatistics was initiated by Ute Vogel. She decisively contributed to my curiosity about the method which I found to be very elegant and powerful. I did not foresee then that it would actually occupy much of my working capacity for years.

Thomas Brinkhoff was the person who opened up the opportunity for me to tackle the problems as covered in this book, which laid the foundation for it to be written in the first place. In the course of my work he was both critical and inspiring and thus often enough paved the way for significant progress.

As my first doctoral supervisor, Manfred Ehlers provided some very valuable hints to significantly improve the quality of my dissertation and therefore of this book too.

With respect to some specific problems of geostatistics, I need to thank Helge Borman and Katharina Hennebőhl for their expertise on the subject and their willingness to contribute to my efforts to create a monitoring simulation framework.

Especially in the final phase of the book project, I drew much benefit from discussions about pursuing ideas with Folkmar Bethmann. His open-mindedness towards complex problems was inspiring to me and helped to advance these approaches.

Apart from the rather specific discipline of geostatistics, I owe many insights about how to organize knowledge in general to extensive reflections about these kinds of problems with Andreas Gollenstede. Motivated by these considerations, I started early in the course of my work to produce a comprehensive taxonomy database about subject matters that were touched. Not only was this approach crucial to grapple with the plethora of themes involved; the basic pattern was also adopted to organize arbitrary complex hierarchies of parameters within the simulation framework.

As a constant source of comfort and encouragement, I owe my wife Christina and our daughter Elisa much of the energy that was necessary to finish the project. This is also the case for Rita, Ulla and Christa.

Contents

Foreword   iv

Preface   viii

Acknowledgements   x

List of Figures   xvi

List of Tables   xviii

1. Introduction   **1**

    1.1 Motivation and Challenges   2
    1.2 Main Contributions   2
    1.3 Observing and Interpolating Continuous Phenomena   9
    1.4 Deterministic Approaches   11
    1.5 Geostatistical Approaches   13
    1.6 Mixed Approaches   15
    1.7 Simulation   16
    1.8 Summary   19

2. Monitoring Continuous Phenomena   **21**

    2.1 Overview   22
    2.2 Requirements   24
        2.2.1 (Near) Real-Time Monitoring   24
        2.2.2 Persistent Storage and Archiving   25
        2.2.3 Retrieval   26
    2.3 Resources and Limitations   27
        2.3.1 Sensor Accuracy   29
        2.3.2 Sampling   29
        2.3.3 Computational Power   31

  2.3.4 Time (Processing and Transmission)   31
  2.3.5 Energy (Processing and Transmission)   32
 2.4 Summary   33

3. Spatio-Temporal Interpolation: Kriging   **37**

 3.1 Method Overview   38
 3.2 The Experimental Variogram   39
 3.3 The Theoretical Variogram and the Covariance Function   40
 3.4 Variants and Parameters   45
 3.5 Kriging Variance   48
 3.6 Summary   50

4. Representation of Continuous Phenomena: Vector and Raster Data 51

 4.1 Overview   52
 4.2 Vector Data Properties   55
 4.3 Raster Data Properties   56
 4.4 Raster-Vector Interoperability   57
 4.5 Summary   60

**5. A Generic System Architecture for Monitoring Continuous   61
   Phenomena**

 5.1 Overview   63
 5.2 Workflow Abstraction Concept   64
      5.2.1 Datasets (Input/Source and Output/Sink)   66
      5.2.2 Process/Transmission   67
 5.3 Monitoring Process Chain   68
      5.3.1 Random Field Generation by Variogram Filter   70
      5.3.2 Sampling and Sampling Density   73
      5.3.3 Experimental Variogram Generation   79
      5.3.4 Experimental Variogram Aggregation   80
      5.3.5 Variogram Fitting   85
      5.3.6 Kriging   88
      5.3.7 Error Assessment   88
 5.4 Performance Improvements for Data Stream Management   89
      5.4.1 Problem Context   90
      5.4.2 Sequential Model Merging Approach   91
          5.4.2.1 Overview   91
          5.4.2.2 Related Work   92
          5.4.2.3 Requirements   92
          5.4.2.4 Principle   93
          5.4.2.5 Partitioning Large Models: Performance   95
                  Considerations

5.4.3 Compression and Progressive Retrieval 98
    5.4.3.1 Overview 98
    5.4.3.2 Related Work 99
    5.4.3.3 Requirements 99
    5.4.3.4 Principle 100
    5.4.3.5 Binary Interval Subdivision 100
    5.4.3.6 Supported Data Types 101
    5.4.3.7 Compression Features 103
5.5 Generic Toolset for Variation and Evaluation of System Configurations 105
  5.5.1 Context and Abstraction 106
  5.5.2 Computational Workload 109
  5.5.3 Systematic Variation of Methods, Parameters and Configurations 113
  5.5.4 Overall Evaluation Concept 115
5.6 Summary 118

6. A General Concept for Higher Level Queries about Continuous Phenomena **119**

6.1 Introduction 120
6.2 Interpolation 121
6.3 Intersection 124
6.4 Aggregation 125
6.5 Conclusions 127

7. Experimental Evaluation **129**

7.1 Minimum Sampling Density Estimator 131
  7.1.1 Experimental Setup 131
  7.1.2 Results 131
  7.1.3 Conclusions 135
7.2 Variogram Fitting 135
  7.2.1 Experimental Setup 136
  7.2.2 Results 139
  7.2.3 Conclusions 141
7.3 Sequential Merging 141
  7.3.1 Experimental Setup 142
  7.3.2 Results 142
  7.3.3 Conclusions 144
7.4 Compression 145
  7.4.1 Experimental Setup 145
  7.4.2 Results 148
  7.4.3 Conclusions 150

7.5 Prediction of Computational Effort   151
    7.5.1 Experimental Setup   151
    7.5.2 Results   152
    7.5.3 Conclusions   152
7.6 Higher Level Queries   153
    7.6.1 Experimental Setup   153
    7.6.2 Results   157
    7.6.3 Conclusions   159
7.7 Case Study: Satellite Temperature Data   162
    7.7.1 Experimental Setup   163
    7.7.2 Results   165
    7.7.3 Conclusions   167

8. Conclusions   **169**

8.1 Subsuming System Overview   170
8.2 Perspective   175

References   **177**

Index   **189**

List of Figures

1.1	Monitoring evaluation principle	7
2.1	Model and monitoring system as mediator	23
2.2	Superordinate monitoring system properties and their interdependencies	34
3.1	Spatio-temporal experimental variogram	39
3.2	Spatio-temporal theoretical variogram	41
3.3	Spatio-temporal covariance function	41
3.4	Different covariance function types	43
3.5	Spatio-temporal variogram models	44
4.1	Raster-vector interoperability in the context of monitoring of continuous phenomena	53
4.2	Raster-vector interoperability	58
4.3	Monitoring interoperability	59
5.1	Monitoring principle for continuous phenomena	63
5.2	Abstraction of a process/transmission step with associated properties	65
5.3	Simulation framework architecture	69
5.4	Pure white noise grid	71
5.5	Random field generation by moving average filter	72
5.6	Nyquist-Shannon sampling theorem	76
5.7	Experimental variogram of sine signal	77
5.8	Experimental variogram point cloud	80
5.9	Binary space partitioning (BSP) process	82
5.10	Aggregation of variogram points using BSP	84
5.11	Weighting functions	86
5.12	Theoretical variogram fitted to aggregated variogram point cloud	87
5.13	Monitoring system architecture	90
5.14	Kriging result with value map and corresponding deviation map	93
5.15	Merging of models by using weight maps	94
5.16	Sequential calculation schema	96

5.17   Theoretical computational complexity of master model calculation   97
       vs. sequential calculation method
5.18   Binary space partitioning (BSP) schema for one dimension   101
5.19   Fencepost error problem for *Integer* values   102
5.1   Binary compression format for progressive sensor data storage   104
5.2   Binary compression format with flexible bit length per dimension   105
5.20   Elements of monitoring taking into consideration the limited   106
       resources, energy, time and storage
5.21   Generic structure to quantify computational effort   110
5.22   UML class diagram for generic organisation of configuration variants   114
6.1   Virtual sensor   122
6.2   Query handler   123
6.3   Polygon-raster intersection   124
7.1   Two-dimensional sine signal as raster grid   131
7.2   Sampling variations applied to a two-dimensional sine signal   132
7.3   Sampling variations applied to a three-dimensional sine signal   133
7.4   Two-dimensional synthetic random field generated by a Gaussian   133
       covariance function
7.5   Sampling variations applied to a two-dimensional random field   134
7.6   Sampling variations applied to a three-dimensional random field   135
7.7   Experimental continuous random field as image sequence   137
7.8   Variogram point cloud aggregation for spatial and   137
       temporal distances
7.9   Separable variogram model fitted to aggregated points   138
7.10   Evaluation diagrams for 108 parameters option variants   140
7.11   Evaluation of sequential method   143
7.12   Performance comparison between master model and sequenced   144
       calculation
7.1   Original header of ARGO drifting buoy data   145
7.2   Header for the compressed dataset of ARGOS drifting   146
       buoy observations
7.3   Compressed data for three observations of ARGO drifting buoys   146
7.4   Compressed data with prolonged bit length   147
7.13   Data volumes (KB) in different formats for 3 datasets   149
7.14   Performance evaluation of four computer system configurations   152
7.15   Aggregation region within continuous random field   155
7.16   Aggregation evaluation with spatial range of 50 m   157
7.17   Aggregation evaluation with spatial range of 30 m   158
7.18   Aggregation process chain   160
7.19   Sea surface temperature (SST) satellite image   163
7.20   Evaluation diagrams for 108 parameters option variants   166
7.21   Variogram generated by the random observations on the sea   167
       surface temperature (SST) image
7.22   SST satellite data interpolation result   167

List of Tables

5.1   Input properties (arranged by main categories: environment,   116
hardware, data, and algorithm) and output indicators of
complex computing systems

7.1   Methodological options for critical steps within the variogram   136
fitting procedure

7.2   Result table with systematic evaluation of best 15 out of   139
108 variogram aggregation variants

7.3   Aggregation indicators   161

7.4   Process method variants for interpolation of sea surface   165
temperature

7.5   Listing of the 15 of 108 configuration variants with the   166
lowest RMSE

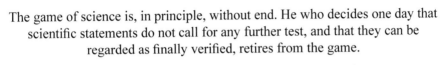

The game of science is, in principle, without end. He who decides one day that scientific statements do not call for any further test, and that they can be regarded as finally verified, retires from the game.

(Karl Popper, The Logic of Scientific Discovery)

Chapter 1

Introduction

CONTENTS

1.1	Motivation and Challenges	2
1.2	Main Contributions	2
1.3	Observing and Interpolating Continuous Phenomena	9
1.4	Deterministic Approaches	11
1.5	Geostatistical Approaches	13
1.6	Mixed Approaches	15
1.7	Simulation	16
1.8	Summary	19

1.1 Motivation and Challenges

Recent developments in the sector of information and communication technology (ICT) have enormously expanded possibilities and reduced costs at the same time [42, 6]. As a consequence, the monitoring of continuous phenomena like temperature or pollution by stationary sensors was intensified since their benefits can be utilized at much lower expenses. There are manifold subject areas which are dealing with phenomena which can be modelled as continuous fields [26, 20]. Analyses based on this specific abstraction model can provide significant benefit to them. The areas of application range from mining, cover matters of geology, oceanology and agriculture and sometimes even touch rather exceptional subjects like medicine or astronomy.

Even more applications can be expected in future because of the universality of the concept of a *continuous field*. Widely differing types and characteristics of phenomena can be incorporated in appropriate *covariance functions* which express the degree of variability of such a field as a function of spatial, temporal or spatio-temporal distance.

Since a field can never be observed as a whole, it has to be estimated from discrete observations by using some interpolation method. Because the correlation of these observations with respect to their spatio-temporal distance is given explicitly, this process of interpolation can be carried out optimally.

In this context, monitoring can be seen as purposeful organization and processing of observations or samples—the terms are used synonymously here—in order to derive a useful model of a particular phenomenon. The main objective is to provide a sufficient estimation of the phenomenon at arbitrary (unobserved) positions (in space and time) at the lowest possible costs.

But in the view of the vast diversity of applications and associated requirements, how should monitoring be carried out for a concrete case? How can a particular phenomenon be characterized and what consequences does this have for the configuration of the sampling and the whole monitoring process? What are the means for reaching well-reasoned decisions in this context? What are the costs of monitoring? Is there a way to continuously increase quality and efficiency of the monitoring process on a systematic basis?

1.2 Main Contributions

Numerous methods can be used to interpolate a continuous field from a set of discrete observations. Generally, there are two principles by which a field can be interpolated from observations: (1) fitting deterministic functions to the observations and (2) assessing their statistical properties and incorporating them into the model. Depending on the phenomenon at hand, a combination of both methods can also be indicated.

This work focuses on the second variant of spatio-temporal statistics or geostatistics (the term is used synonymously here), since its methods are widely accepted and applied and in many cases provide the best results when dealing with continuous phenomena.

The main objectives of this work are (1) to estimate the sufficient sampling density for a given phenomenon and (2) to test different variants of methods and parameter settings of the interpolation. A framework for systematic variation of these factors according to several performance indicators is introduced in order to evaluate them. It is designed for continuous improvement of the overall efficiency of the monitoring process.

In the context of monitoring continuous phenomena there are many challenges concerning the associated tasks of observation, transmission, processing, provision and archiving. There has been and there is continuous progress with respect to increasing hardware performance and decreasing costs. Also, the algorithms associated with monitoring become more powerful and matured.

There are many studies about the processing of concrete datasets of sensor observations in order to derive a continuous field. There is also a vast number of works dealing with the theoretical foundations of geostatistics, although the consideration of spatio-*temporal* modelling is still not very common in this context [31, 52].

What is missing in the author's viewpoint is a systematic examination and evaluation of the process of monitoring as a whole. The intention of the framework introduced here is to support an iterative calibration of the used process model. Since diverse performance indicators are provided with each variant of a simulation, the whole process chain can be regarded as a "closed loop" where input data and parameters can be related to the output quality [116, p. 9]. Continuous learning about and improvement of the monitoring process is thus facilitated [16].

The framework presented in this work covers the entire workflow of a simulated monitoring using kriging as the interpolation method. Each step of this workflow is listed below. The specific contribution of this work if present is added to each particular step.

1. **Random field generator** The central theme of this work is the investigation of environmental phenomena which can be regarded as continuous in space, time or space-time like temperature, air pollution, radiation and other variables. The strategy for sampling has to consider the dynamism of the phenomenon and at which level of detail this dynamism has to be captured. So one fundamental question for sufficient observation is how dynamism is related to the minimum sampling density that is necessary to capture it adequately.

The spatio-temporal dynamism of a continuous random field is controlled by the moving average filter that is used to generate it. All subsequent process steps, starting with sampling, can be tested against changed initial conditions according to this dynamism. The generality of the applied models and approaches can thus be corroborated [47, p. 62].

2. **Sampling** The critical nexus between the phenomenon itself and its model is established by sampling. The density of the sampling determines at which granularity level the phenomenon is captured. Too sparse sampling can never yield the true character of the observed phenomenon no matter how sophisticated the interpolation is. The geostatistical parameter *range* is an indicator for the dynamism (in space, time or space-time) and therefore also determines the minimum necessary sampling density.

   In this work, a formula is deduced from signal processing that estimates the minimum necessary sampling density from the given *range* parameter, which expresses the maximum distance of autocorrelation. The approach involves experimental evaluation. It provides an objective estimation of the necessary sampling density for a given phenomenon and thus makes different observational settings comparable in principle.

   Another important issue concerning observational data is its efficient transmission and archiving. A compression algorithm is proposed that is designed for this data structure and capable of progressive retrieval.

3. **Experimental variogram** The experimental variogram expresses how sensor observations are actually correlated with respect to their spatial, temporal or spatio-temporal distances. For each possible pair of observations, this distance is related to the corresponding semi-variance, which is the halved and squared difference between the measured values. A plot of this relation already conveys an impression of the statistical behaviour of the observed variable with respect to correlation that depends on spatial, temporal or spatio-temporal proximity. The experimental variogram is a prerequisite for subsequent geostatistical analysis.

4. **Aggregation of the experimental variogram** To be applicable for interpolation by kriging, the experimental variogram generated by the previous step needs to be represented as a mathematical function. The parameters of this function are fitted to the empirical data. Since the number of points of the experimental variogram grows by $\frac{n^2-n}{2}$ for n observations, the fitting procedure can become expensive even for moderate amounts of data.

   The aggregation of variogram points is one approach to cope with this problem. Such aggregation is usually carried out by a regular partitioning of the region populated by points of the experimental variogram. In this work, the process is carried out with respect to the statistical properties of the point set that is to be partitioned. Different variations of this approach are tested.

5. **Fitting of the theoretical variogram function** With only the aggregated points instead of all variogram points, the fitting procedure can be executed with much less computational effort. The Gauss-Newton algorithm is often used to minimize the residuals of the aggregated points according to the function by adjusting its parameters iteratively [116, 113]. By introducing weights, the points representing low distances can be given more influence, which is a reasonable strategy here because bigger distances also tend to be less reliable for parameter estimation due to higher dispersion. Different weighting strategies are tested and evaluated in this work.

   In order to make the estimation of optimal parameters by the Gauss-Newton algorithm more robust, starting values for the optimization are deliberately chosen from an n-dimensional grid within quantile borders of each dimension. This alleviates situations where the Gauss-Newton algorithm does not converge or finds several local minima.

6. **Interpolation by kriging** Given the parameters as derived from the variogram fitting, the interpolation can be performed at arbitrary positions and therefore also for arbitrary grid resolutions to fill spaces between observations. As a statistical method, kriging represents unbiased estimation of minimum variance [29, 129]. Beside the value itself, kriging also provides the estimation variance derived from its position relative to the observations it is interpolated from [89, p. 464], [97].

   The kriging variance is a unique feature and can be exploited for several purposes. In this work it is used as weighting pattern when merging several raster grid models. The computational effort for kriging can be reduced when subsets of observations are processed and merged sequentially. Merging can also be used to seamlessly integrate new observations into existing models. This is crucial when a (near) real-time model of the phenomenon has to be provided.

7. **Higher level queries** Statements about continuous phenomena like an exceeded threshold of pollutants of a particular region within a particular period of time can not easily be drawn from raw sensor data. There is much ambiguity about how to process the available observations in order to derive the best knowledge possible to either confirm or refute such a statement. Based on the strong expressiveness of the kriging variance, this work proposes a concise method to quantify the reliability of such statements.

8. **Performance assessment** Because the synthetic reference model that is observed can be created at arbitrary resolution, the deviation of a model derived from sampling and interpolation can be calculated exactly. When this derived model is provided at the same extent and resolution as the reference model, the root mean square error (RMSE) can easily be calculated easily. This value is the key indicator for the simulation because it expresses the overall quality of the monitoring process [51]. Algorithmic variants and parameter adjustments will affect this value and can therefore be used for iterative optimization [47]. Other performance indicators like computational effort can also be improved this way.

The operational steps listed above constitute the components of a monitoring environment that derives raster grid models from discrete observations of continuous phenomena. It is designed to systematically vary and evaluate different methods and parameter settings of monitoring in order to iteratively increase efficiency.

Efficiency in this context can be defined as the relation between the expenses that are necessary to operate a monitoring system and the quality of the model as derived from observation and interpolation.

In order to express and systematically evaluate this efficiency, the following aspects of a monitoring scenario need to be quantified:

■ extent and dynamism of the phenomenon
■ sampling effort
■ computational effort
■ model quality

These issues above are interdependent. When planning a monitoring system, the first task is to define the extent and to estimate the dynamism of the phenomenon to be observed. The second task is to decide about the necessary granularity and accuracy of the model to be created by the monitoring. Given adequate knowledge about these two conditions, the monitoring system should be designed to sufficiently mediate between them [13, p. 6].

It is up to the decision makers to choose the hardware and software that is appropriate under the given circumstances. The present work is intended to provide methods and tools to support this aim with approaches that can be corroborated experimentally. Continuous efficiency gain concerning the ratio between used resources and achieved accuracy can thus be facilitated. Following the idea of a *closed loop* as also propagated by Sun & Sun on [116, p. 9], and the guiding principle of this work is depicted in Figure 1.1.

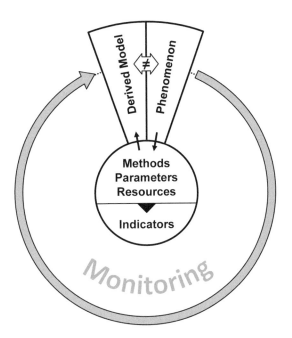

Figure 1.1: General principle of evaluation of monitoring.

The schema basically represents the two leading policies that are applied in this work to foster methodological improvement in monitoring continuous phenomena:

1. **Circularity of the monitoring process:** The simulated sampling is carried out on the synthetic reference model. The derived model is generated by kriging interpolation of these samples. When the reference model is given with arbitrary accuracy, as is the case for a synthetic model, the deviation between the derived model and the reference model can be determined exactly.

2. **Systematic variation of methods and parameters and evaluation of output indicators:** The monitoring process as a whole can be configured by various methods and parameters. Namely, these are the density and distribution of observations, the applied interpolation algorithms with their associated settings and also the computational resources used. Variations of these factors will more or less affect the output indicators.

Given this generic framework, continuous improvement of the applied methods and parameters can be fostered. Quantitative evaluation of the monitoring quality and efficiency can be carried out by appropriate indicators for accuracy and computational effort. In combination with a model of the available computing resources, this effort can be concretised in terms of time, energy and storage.

These are crucial constraints especially for large models, (near) real-time systems and wireless sensor networks and should be considered carefully.

The main challenge of this work is to furnish this general framework with methods, parameters and indicators that are appropriate to optimise the task of monitoring continuous phenomena given limited observations and resources.

The remainder of this book is structured as follows:

The properties of continuous phenomena which are the subject of investigation of this work, are characterized in the remaining sections of this chapter. A general overview of common interpolation methods is also given here.

Beyond observation and interpolation of such phenomena, the process of monitoring entails many technical and organizational issues that need to be considered in a real-world scenario. These will be covered in Chapter 2.

The statistical interpolation method used in this work, namely kriging, is described in more detail in Chapter 3.

When representing continuous phenomena on a computer, the most fundamental categorization is that of raster data vs. vector data. Both formats and their interoperability in this context are outlined in Chapter 4.

On this basis, a system architecture for monitoring continuous phenomena is presented in Chapter 5. It addresses the problems worked out in the previous chapters and is designed to systematically and iteratively improve the efficiency of the monitoring process as a whole.

Being the actual rationale behind monitoring, the determination about the presence or absence of complex states is addressed in Chapter 6. A framework based on the abstraction of the monitoring process identifies generic building blocks of a monitoring system architecture that aims at a high degree of interoperability.

An experimental evaluation of the proposed concepts is carried out in Chapter 7 before conclusions are drawn in Chapter 8. A general perspective on the future development of monitoring systems is also sketched out here.

1.3 Observing and Interpolating Continuous Phenomena

Continuous fields can serve as an appropriate model to describe a variety of phenomena. In fact, most environmental variables are continuous [129, p. 57]. Therefore, methods and tools to handle continuous fields are common in many subject areas [30, p. 11], [8, p. 1], [83], [21]. Some of them are listed below without any claim of exhaustiveness:

- agriculture and soil science
- astronomy
- climatology and meteorology
- ecology (flora and fauna)
- economics
- environmental science
- fishery
- forestry
- geology
- hydrology and hydrogeology
- medicine
- mining and petroleum engineering
- pollution control
- public health
- remote sensing
- social geography

The reason for this diversity is that the principle of a continuous field is so universal [68, 25]. Yet, the methods for handling observational data about these phenomena are still evolving. Unlike imagery data where the end-product is usually of similar resolution as the observation itself—e.g. when carried out by CCD sensors—, the sensor data that provide discrete values at particular positions in space and time have to be handled differently [20, 76, 25]. Although the resulting data type—for an easy interpretation by humans as well as by machines—may in fact be discretised as raster grid, this does not at all imply an analogy of the acquired data.

For sensor data registered by regional stations there is a substantial gap to bridge between the original observations and the format required for reasonable interpretation or analysis. This has several impacts on the way such data need to be treated in terms of accuracy, coverage and interoperability:

Accuracy

When dealing with grid data from remote sensing, errors can be caused by the sensor itself, by atmospheric effects or by signal noise [88]. There might also be effects to consider caused by pre-processing the data or resampling it to a different resolution in space and time. Excluding systematic trends, the accuracy of the raster cells is more or less homogeneous. In contrast to that, the confidence interval will vary significantly for a grid derived from interpolated observations of regional stations. It depends on the distances of each interpolated grid cell to the observations surrounding it.

Coverage

When a region is to be monitored, not only its extent, but also the observational density has to be considered for both space and time. Unlike for remote sensing [110], where ground resolution is already an intuitive metric, for interpolation it is necessary to relate the dynamism of the phenomenon to the sampling density (see Section 5.3.2) in combination with the quality and appropriateness of the method and its parameters (see Section 5.3.4 and 5.3.5).

Interoperability

For visualization and analysis that involves other spatio-temporal referenced data, appropriate formats and interfaces have to be provided to access the field data. In this context, a high level of abstraction is a prerequisite for interoperability [132, p. 30]. Querying a variable at arbitrary positions in space and time can be seen as the most basic function here [27]. It can easily be extended to a grid-based structure which directly supports visualization and analysis. On a more sophisticated level, regional maxima or average values or other types of aggregation can be provided. A general concept for the definition of such queries should precede a syntactical specification of formats or interfaces.

Given these properties of interpolated data, it might appear reasonable to prefer imaging techniques like remote sensing to stationary observations. A coverage of the area of interest by a raster grid that directly reflects the acquiring method and provides homogeneous accuracy is certainly advantageous. Unfortunately, such observations are often unavailable, too expensive or just not applicable to the particular problem. For these situations, interpolation is the only way to provide a gapless representation of the sparsely observed phenomenon. This is the method this work focuses on.

For environmental monitoring, variables like wind speed, precipitation, temperature or atmospheric pressure can be observed by weather stations. The sparsity of the observations according to space and time necessitates reasonable estimations of the value at unobserved positions. There are two general principles for interpolation to provide them: determinism and statistics [2, 60, 129]. Deterministic approaches align parameters of mathematical functions to initial conditions while statistical ones assume the observed phenomenon as a result of a random process that is autocorrelated according to spatial, temporal or spatio-temporal distances of observations.

Regardless of the method used for interpolation, the observational data itself requires particular structure to make it valuable for interpretation. Whittier et al. suggest a space-time cuboid on which sliding window queries can be performed in [131]. A spatio-temporal interpolation is performed for each cell that is not covered by an observation. The structure resulting from this process can be interpreted as a three-dimensional grid or *movie*.

A more abstract approach is introduced by Liang et al. [76]. A specific data type to manage observational data about continuous phenomena is defined at a conceptual level. The general idea is to store observational data in a standardized way and to select the interpolation method when retrieving the data. So instead of generating and storing interpolated data as additional grid dataset that has to be managed separately, the method integrates observations, interpolation and derived data to one coherent model. Thus, derived grids might be generated immediately with new observations or just on demand when the region is queried. Also mixed strategies are possible here since the management of the data can remain totally transparent to the user or application when using an appropriate query language for field data types.

The vision of such an integrated mechanism for field data is yet far from interoperable realization in available systems, although it appears to be a superior concept. There is still much effort necessary in terms of standardisation of naming and implementation of interpolation methods. At least the most frequently used interpolation methods should simply be called by query parameters and provide identical results from different systems. Since spatio-temporal interpolation is a very complex process with an immense variety of methods, variants and parameters [75] that is still continuously growing and evolving, is be a challenging objective. Nevertheless, in the long run it will be necessary to delegate this specific task to a basic infrastructure service component (e.g. a data stream engine [42]) to unburden higher-level applications from this complexity. A similar development has taken place for geometries that are stored in database management systems [17].

Without claiming completeness, a list of commonly used interpolation methods is provided in the next two sections which are named by the most general classification dichotomy: deterministic vs. statistical, or rather, to be more specific, geostatistical methods. Approaches that combine both principles will be covered briefly in the subsequent section.

1.4 Deterministic Approaches

There are various interpolation methods that do not take into account the random character of the observed field and are therefore classified as deterministic [129, 75]:

Voronoi polygons

or Thiessen polygons tessellate a region into polygons so that for each position within a polygon one particular observation is the nearest one. All these positions share the exact value of that observation. As a consequence, there are sudden value steps at the borders between these polygons. Such discontinuities restrict the scope of application.

Triangulation

provides a surface without discontinuities or "jumps" of value by filling the space between three observational points with tilted triangular plates. It has, though, abrupt changes in gradient/slope at the triangle edges.

Natural neighbour interpolation

is based on Voronoi polygons. It extends the concept by introducing weights that are proportional to the intersection areas between the Voronoi polygon of the point to be interpolated and the ones of the neighbouring points. In contrast to the preceding approaches, it provides a continuous surface.

Inverse distance

inverse distance weighting presumes that the influence of an observation on the interpolation point is decreasing with increasing distance. This decrease is expressed as the inverse of the distance with an exponent bigger than zero.

Trend surfaces

defined by mathematical functions are another way to represent continuous fields. The functions' parameters are fitted to the observations by regression. With an increasing number of observations this approach becomes numerically fragile and the residuals at the observed positions tend to be autocorrelated.

Splines

can also be used to create continuous surfaces. They are based on polynomial functions, but there are multiple instances of them that are join smoothly by fitting their parameters, positions and orientations.

In summary, deterministic in the context of interpolation means that there is a particular law by which the continuum of a value can be *determined*. Just as phenomena that are described by Newton's physical laws, there is no consideration of randomness [102]. The parameters of these deterministic laws or functions are fitted to actual data, but randomness is not incorporated into the interpolation method. An estimation of variance for the interpolated value can thus not be provided.

Deterministic methods only rarely represent the nature of the environmental phenomenon in a sufficient way. There are usually many complex physical processes involved to produce the particular phenomenon [129, p. 47]. Because it is impossible to keep track of all of them, it is often reasonable to regard them as one random process [60, p. 196 ff.]. This approach is discussed in the next section.

1.5 Geostatistical Approaches

In contrast to deterministic methods, geostatistical methods *do* take into account the stochastic nature of the phenomenon at hand. The geostatistical method of *kriging* determines the interpolation value of minimum variance with respect to the covariance structure as expressed by the variogram. It should be the first choice wherever the observed phenomenon can be considered stationary random process, at least approximately.

Stationarity means that the statistical properties of a process are invariant to translation [31, p. 34]. While *first-order* or *strong* stationarity implies that *all* statistical moments remain constant, *second order* or *weak* stationarity only encloses mean, variance and the covariance function. *Intrinsic* stationarity reduces the conditions to the consistency of the variogram with the data [129, p. 268 f.]. Actually, the interpolation of intrinsic phenomena can be carried out using the same kriging system [8, p. 90]. Strong stationarity is rather a matter of theory and even weak stationarity is not a prerequisite for kriging in practice [31, p. 323].

Hence, formal geostatistical concepts like stationarity should not be overestimated according to their practical value. Real world conditions only rarely satisfy theoretical considerations and any model "can he considered false if examined in sufficient detail", as Beven [13, p. 38] points out.

Nevertheless, with its wide range of variants and parameters kriging provides a sophisticated toolset to adapt to a large variety of phenomena. Within geostatistics, kriging is the most important method, or, as Cressie writes in [29, p. 239]:

> The use of the word "kriging" in spatial statistics has come to be synonymous with "optimally predicting" or "optimal prediction" in space, using observations taken at known nearby locations.

Or, as Appice [6, p. 51] puts it:

> [...] kriging is based on the statistical properties of the random field and, hence, is expected to be more accurate regarding the general characteristics of the observations and the efficacy of the model.

Its superiority compared to other methods is also emphasized by de Smet [33]:

> Of the studies that intercompared methodologies (Bytnerowicz et al. 2002), kriging was objectively shown to give the best results.

The expressive power of kriging has also made it popular in machine learning, where a generalization of the method is known as *Gaussian process regression* [106, p. 30], [116, p. 351], [45].

In contrast to deterministic approaches, geostatistical methods incorporate the random nature of a phenomenon by introducing the concept of the *regionalized variable*, which is characterized as:

$$Z(x) = m(x) + \varepsilon'(x) + \varepsilon''(x), \qquad (1.1)$$

where $m(x)$ represents the structural component or trend, $\varepsilon'(x)$ is the autocorrelated random term and $\varepsilon''(x)$ is the uncorrelated random noise. [19, p. 172].

Kriging exploits the character of the stationary variable to provide unbiased estimations of minimum variance [129, 31].

As already mentioned, stationarity, or more precisely, second-order stationarity, implies constant mean, variance and covariance function, or, as Lantuejoul puts it more concretely in [70, p. 24]:

- there is a finite mean m independent of x
- the covariance between each pair is finite and only depends on the pair's distance

The covariance function is thus specified as the central geostatistical concept by which the variance within pairs of values is expressed as a function of distance. In most cases, this correlation decreases with increasing distance. This explicit rule of autocorrelation is applied when the degree of contribution of each single observation to an interpolation is estimated by an optimal weight. The optimal weight estimation itself is a linear regression problem [95] with the associated solution of matrix inversion and therefore of comlexity $\mathcal{O}(n^3)$ [24, 10], [46, p. 503].

Whereas the optimal weight estimation is influenced by the observational values themselves, the *kriging variance* is only determined by the covariate structure expressed as covariance matrix at the interpolation point [54, 45]. It expresses the degree of uncertainty or variance that can be expected from the relative positioning of the interpolation point towards the observational points used for interpolation. This *kriging variance* is a crucial information in the context of a setting where (spatio-temporal) autocorrelation is empirically investigated and expressed by a covariance function.

Besides the estimation of uncertainty at a particular position, the kriging variance can be used for sampling configuration and adaptive sampling [126, 54, 45]. In this work, the kriging variance is used to merge sub-models in order to improve performance or to provide a continuous update mechanism for (near) real-time environments (see Section 5.4.2). Beside that, it is used as confidence measure when performing higher level queries (see Chapter 6).

Notwithstanding the sheer overwhelming variety of kriging variants, this work sticks with the basic version of the method known as *simple kriging* [31].

Furthermore, neither noise (*nugget effect*, see [31, p. 123], [129, p. 81]) nor variation of semivariance with direction (*anisotropy*, see [19, p. 181], [31, p. 128] are considered in favour of the more general aim of systematic variation and evaluation of methods and parameters. However, the still rather exceptional aspect of *temporal* dynamism of the phenomenon [31, 53, 52] is covered by applying the associated spatio-temporal covariance functions (see Section 3.3).

1.6   Mixed Approaches

On a conceptional level, the dichotomy between deterministic and stochastic methodologies is helpful for a thorough understanding of different approaches. In practice, however, the observed phenomena appear as manifestations of both principles, as Agterberg points out in [2, p. 313] for the realm of geology:

> It is important to keep in mind that trend surfaces in geology with residuals that are mutually uncorrelated occur only rarely. More commonly, a variable subject to spatial variability has both random (or stochastic) and deterministic components. Until recently, there were two principle methods of approach to spatial variability. One consisted of fitting deterministic functions (as developed by Krumbein and Whitten), and the other one made use of stationary random functions (Matheron and Krige).

In geostatistics, this issue is addressed today by modelling a deterministic trend, as is the case with universal kriging [120], [19, p. 186], [95, p. 85].

A fusion of deterministic and statistic approaches appears to be a general trend, as Chiles & Delfiner point out in [21, p. 10]: "The current trend in geostatistics is precisely an attempt to include physical equations and model-specific constraints." Likewise, Poulton points out in [103, p. 192]: "Practical methods may be the joint application of deterministic and statistical approaches."

Generally, the distinction between deterministic and stochastic effects is one of the most fundamental problems of science [102]. In the context of environmental monitoring, however, it is not of decisive importance whether a particular phenomenon is predominantly seen as the result of deterministic or stochastic processes. Rather, the monitoring process should be evaluated by the quality of the model it derives based on the available observations.

A simulation environment is a powerful tool to systematically perform such evaluation because it provides full control over the phenomenon model, the sampling and the parameters, and principally unlimited knowledge about the quality and efficiency of the simulated monitoring process, as will be outlined in the next section.

1.7 Simulation

Depending on the subject area, the term *simulation* can have different meanings and therefore different prerequisites. As Pritsker [105, p. 31] points out, a simulation is based on a *model*, which is an abstracted and simplified representation of the system under investigation. Predicting the dynamic behaviour of such a model given its initial conditions is then called *simulation*. Likewise, Birta & Arbez [14, p. 3] identify the central characteristic of "behaviour over time" or, as Bank [9, p. 3] puts it: "[the] imitation of the operation of a real-world process or system over time."

Simulation has a wide area of application. Wherever a system is too complex to be described analytically—which is the case for most systems of interest—, its behaviour can be simulated given the laws and initial conditions. Depending on the goal of the modelling and simulation, a system can be inspected on different levels of knowledge and complexity [132, p. 13], [69, p. 10].

The scope of modelling and simulation is by far wider than the one covered by this work. It can be applied to investigate problems of production, healthcare, military, customer behaviour, traffic, to name just a few [73, 9, 132]. The focus here, however, lies on environmental phenomena considered to be continuous in space and time. A thorough reflection of the role of modelling and simulation in this context is given by Peng et al. [99, p. 9]:

> A well-tested model can be a good representation of the environment as a whole, its dynamics and its responses to possible external changes. It can be used as a virtual laboratory in which environmental phenomena can be reproduced, examined and controlled through numerical experiments. Environmental models also provide the framework for integrating the knowledge, evaluating the progress in understanding and creating new scientific concepts. Most importantly, environmental modelling provides the foundation for environmental prediction. Environmental models are useful for testing hypotheses, designing field experiments and developing scenarios.

In this book, the term *simulation* can be applied on two major issues of the monitoring scenario:

1. The continuous random field representing the spatio-temporal dynamism of the phenomenon
2. The density and distribution of observations carried out on that random field

It could be argued that for purely spatial random fields and the associated observations there is no dynamism at all. Therefore, such a scenario can hardly be called a simulation. On the other hand, the consideration of the temporal dimension would fulfil the prerequisite for a simulation while it would not change the monitoring process *in principle* but just bring in one more dimension. Furthermore, unlike for Monte Carlo methods [108] which can be assigned to the domain of numerical analysis rather than simulation [14, p. 13], continuous random fields are generated to represent real phenomena instead of pure mathematical models. They simulate processes like sedimentation, erosion, diffusion and the like, which in effect are so complex that they can be seen as stationary random.

A more general classification schema for simulation methods is given by a Law [73]. It allows for categorization by the following three dimensions:

- static vs. dynamic
- deterministic vs. stochastic
- continuous vs. discrete

Aral [7, p. 44 ff.] adds the linear vs. nonlinear to this list of dichotomies, whereas linearity is only applicable to very simple models. Jorgensen [64, p. 28 ff.] adds some more elements which are not considered relevant here.

The classification of the simulations carried out in this work is not without ambiguities when using such schemata. So the synthetic continuous random field can be seen as static in time, but only when the spatial dimension is excluded. While the field can be seen as continuous—although discretised to a raster grid—the observations, being part of the simulation, are discrete events in space-time. These events can either be carried out deterministically by following an observation plan or stochastically by scattering them randomly in space and time. So it can be said that the concepts and categorizations for general simulation do not necessarily apply to the realm of continuous environmental phenomena.

When shifting to the domain of geostatistics, continuous fields are regarded as outcomes of random processes [30]. Describing such a process by a statistical model and running it with a computer is actually called *simulation* [70], or, as Webster & Oliver [129, p. 268] put it:

> In geostatistics the term 'simulation' is used to mean the creation of values of one or more variables that emulate the general characteristics of those we observe in the real world.

There are generally two variants of simulating continuous fields: unconditional and conditional [70, 129]. When carrying out an unconditional simulation, the main interest is to create a random field with properties of a particular covariance function. No further constraints are laid on the realization.

In contrast to that, the idea behind conditional simulation is to create such a random field through a set of real or fictitious observations. These observations keep their values in the simulated realization whereas for the positions between them, random values will be generated with respect to the associated covariance function [70].

Alternatively, one could also think of just applying kriging interpolation to those observations. But while kriging provides estimates of no bias and minimal variance, the dispersion of the phenomenon is not necessarily represented by it [129, p. 271]. So the simulation is to be preferred to interpolation when the overall statistical character of a field is more important than the best possible estimation (no bias, minimal variance) at each position.

Technically, there are several methods to create such random fields. Probably the most popular is the lower-upper (LU) decomposition of the covariance matrix. It has the disadvantage that for n grid cells there is a matrix of dimension $n^2 \times n^2$ to be decomposed. This can exceed computational capacities even for moderate model sizes.

Beside this method, Webster & Oliver [129] entitle sequential Gaussian simulation, simulated annealing and turning bands as simulation techniques.

Lantuéjoul [70] is listing dilution, tessellation, spectralisation and turning bands as methods for generating continuous random fields.

The approaches above are either limited in their field of application or lack cohesion between the generated field and the covariance function.

In contrast to that, the moving average filter provides a flexible and intuitive way to generate a random field representing a particular covariance model. The concept is analogous to spatial filtering in signal processing [49, 115]. The filter is applied to a field of independent and identically distributed values (pure Gaussian noise). The output value of each grid cell is a weighted average of the corresponding cell and its surrounding cells in the input grid, whereas the weight decreases with increasing distance from the target grid cell. The process is repeated for each grid cell and can be imagined as a moving filter, mask, kernel, template or window [49, p. 116]. The weighting scheme of such a kernel determines the autocorrelation structure of the output random field and can be derived from an appropriate covariance function (see Section 5.3.1).

Oliver [94] analytically derives kernels for moving average filters for the most common covariance functions (spherical, exponential, gaussian). When applied as filters on Gaussian random fields, these kernel functions produce continuous random fields which are compliant with the covariance functions the kernels were derived for.

In this work, however, there is no rigorous mathematical derivation of the covariance functions used to define the kernel of the filter. Consequently, the continuous random fields which are generated using these functions as kernel filters do not fulfil the conditions of stationarity in the strict sense. Due to the generation process they are, however, random, spatio-temporally autocorrelated and isotropic.

The relationship between the covariance function of the kernel filter and that of the resulting field is analytically demanding [94] and beyond the scope of this work. But since real world phenomena do not obey formal statistical considerations either, we neglect this rigor here and focus on the methodological approach for continuous improvement of the monitoring of continuous phenomena as a whole.

1.8 Summary

Continuous phenomena are ubiquitous and their monitoring and analysis are common tasks for many disciplines. The main challenge is to choose sampling schema and interpolation method in order to generate an appropriate model of the phenomenon. For many natural phenomena there is continuity in both space and time. The dynamism in each of these dimensions should be captured according to the monitoring objectives.

There are manifold methods to interpolate between discrete observations of a continuous phenomenon. The geostatistical method of kriging considers the observed phenomena as a random process with a particular autocorrelation structure. Being a powerful method of high adaptivity, it can incorporate complex correlation structures, deterministic trends as well as anomalies like anisotropy. One key feature of kriging is the estimation of uncertainty. It is exploited in this work for sequential merging of sub-models and to assess the reliability of statements derived from aggregations.

Regardless of the interpolation method that is applied, it is helpful to consider the idea of a field data type as a general abstraction concept [20, 76, 26, 25]. A monitoring system can thus be seen to mediate between *discrete* sensor observations and a *continuous* field that represents the phenomenon of interest. The discrepancy between this phenomenon and the derived field representation expresses the quality of a monitoring process. With the help of a synthetic continuous random field this performance metric can be used to guide continuous improvement of quality and efficiency of the whole monitoring process.

Chapter 2

Monitoring Continuous Phenomena

CONTENTS

2.1	Overview	....	22
2.2	Requirements	....	24
	2.2.1	(Near) Real-Time Monitoring	24
	2.2.2	Persistent Storage and Archiving	25
	2.2.3	Retrieval	26
2.3	Resources and Limitations	....	27
	2.3.1	Sensor Accuracy	29
	2.3.2	Sampling	29
	2.3.3	Computational Power	31
	2.3.4	Time (Processing and Transmission)	31
	2.3.5	Energy (Processing and Transmission)	32
2.4	Summary	....	33

2.1 Overview

The monitoring of continuous phenomena like temperature, rainfall, air pollution and others, is carried out by (wireless) sensor networks today. From the perspective of the user of such data, the original discrete sensor observations are not very useful since they often do not cover the actual spatio-temporal area of interest. So the fundamental task in this context is to provide a gapless and continuous representation of the particular phenomenon either as a visualisation in real time or as a model for long term archiving, or both. Therefore, it is necessary to cover the area of interest with observations in a way that is sufficient to capture the phenomenon (see Section 5.3.2). The necessary density of observations depends on the dynamism of the phenomenon; for spatio-temporal monitoring this has to be considered for both space and time. From these observations, the value of the particular phenomenon needs to be estimated for unsampled spatio-temporal positions.

Monitoring as a whole can be seen as an optimization problem or trade-off between spent resources and achieved model quality, or, as Cressie [31, p. 26] puts it:

> Looking at this from another angle, the best scientists collect the best data to build the best (conditional-probability) models to make the most precise inferences in the shortest amount of time. In reality, compromises at every stage may be needed, and we could add that the best scientists make the best compromises!

Notwithstanding the fact that sensors and computers become cheaper and more efficient, resources will always remain limited, which constraints the process of monitoring as a whole to be as efficient as possible.

Essentially, each monitoring system can be seen as a mechanism that uses observations to provide knowledge about the environment that is required by society. An abstract model of the phenomenon of interest and a monitoring system based on it are the components which mediate between environment and society [37].

Figure 2.1 illustrates these very general components of environmental monitoring (environment, model, system, society) with their respective relations and interactions (validation, verification, credibility, monitoring, science, observation, knowledge). Validation, verification and credibility are established concepts in the field of simulation and modelling (see [73, p. 246 ff.], also [9, 132]).

Whereas validation is necessary for choosing the right model, the verification explores if it was implemented correctly. The credibility expresses the degree of acceptance of a particular system solution among its users, consumers or other stakeholders, while science rather discusses and improves the system-independent concepts of the underlying models. Validation as its component contributes to this process.

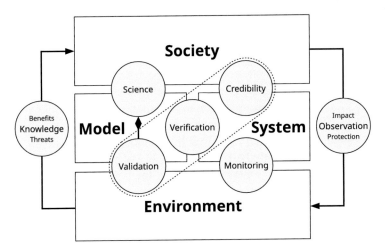

Figure 2.1: Model and monitoring system as mediator between environment and society; validation, verification and credibility (inside dashed frame) as established concepts from simulation and modelling, monitoring as data feeding process, and science as social process developing and validating abstract models. Observation is organized by society in order to gain knowledge, exploit benefits, avert threats, and protect the environment from hazardous impacts.

The superordinate goal of this arrangement is to endow the society with knowledge about the environment to better exploit its resources (benefits) and be protected from potentially hazardous processes (threats). In order to gain such knowledge, the society organizes observations that continuously feed the system with data (monitoring). This knowledge can be used to identify adverse human impact and induce protection policies in a fact-based manner.

This book focuses on continuous environmental phenomena (e.g. temperature, humidity, air pollution) and provides methods and tools that facilitate the iterative development of effective and efficient monitoring scenarios.

On a conceptional or interim level, a continuous phenomenon can be represented abstractly as a field. In order to be easily interpreted by both humans and information systems, the field needs to be discretised to a raster grid of appropriate resolution [13, p. 41], [26], which then becomes the carrier of information about the phenomenon (see Chapter 4).

Given that the phenomenon—at least in principle—fulfils the criteria of stationarity [129, 31, 124], the geostatistical method of kriging, among spatio-temporal interpolation methods, is often the best choice to derive such a continuous representation from discrete sensor observations (see Chapter 2).

However, to generate such a continuous representation of a phenomenon is a complex calculation process with distinct stages and associated intermediate results. Each of these stages entails several algorithmic variants and parameters that control the process. In a real monitoring environment, the optimal methods and parameters for this process usually remain unknown and can only be estimated by ad hoc heuristics.

In contrast to that, a synthetic continuous random field provides the exact state of the phenomenon at any position in space and time. Therefore, the resulting accuracy of any monitoring process (sampling and interpolation) can always be quantified by the difference between the simulation model and the model derived from the interpolated observations.

The simulation framework described in this book was designed to systematically test a wide range of algorithmic variants and parameter settings and inspect their effects on several indicators. Beside the deviation from the reference model, also the actual computational effort that is necessary for each variant can be considered. This makes it possible to quantify and thus compare the efficiency of different approaches. By abstracting the computational effort of a particular calculation from the hardware, it is used in principle possible to estimate the expenses incurred in terms of time and energy for any other given computer platform. This can be a critical aspect for wireless sensor networks, large models and environments with real time requirements.

2.2   Requirements

Before inspecting its efficiency indicators, the general requirements of a monitoring system will be specified in the next subsections.

2.2.1   (Near) Real-Time Monitoring

For many applications it is crucial that the model derived from monitoring is provided in real time or near real time. This is especially the case when the observed phenomenon potentially has severe impacts on security or health like radiation, pollution or heavy rainfall [7]. In all these cases it has to be ensured that observation, data transmission, generation and provisioning of the model is carried out in time according to the requirements.

When the model derived from monitoring is to be provided in real time—e.g. via a web mapping application—one major problem is how to continuously update it. This is especially problematic where observations are irregularly scattered in space and time, as is the case for autonomous mobile sensor platforms. In a sensor data stream environment, one might initially consider two ways to cope with this problem:

1. Appending the new observations to the existing set of observations and calculating the model anew
2. Creating subsequent subsets of particular size (e.g. 10 minutes time slices) with separate disjunct models per subset

The problem with the first solution is that the number of observations will soon overstrain the computational capacities necessary for immediate model calculation. The problem with the second solution is to choose an appropriate size of the time slice: very short intervals will lead to deficient models due to data sparsity. Very long intervals will be a burden on model calculations on the one hand and will undermine the actuality of the provided model on the other.

Another approach is to not completely replace the previous model by the one generated from the latest set of observations, but instead *merge* the new model with the previous one. The kriging variance that is—beside the value itself—available for each grid cell of the model can be used for weighting when merging two grid models (see Section 5.4.2). A continuous and flexible update mechanism which is necessary for timely monitoring is thus provided.

The merging algorithm also addresses another requirement of real-time monitoring: coping with heavy computational workload when kriging large datasets of observations. When numerous observations have to be processed in near real time, this can become a critical factor even for powerful computing environments. Therefore, appropriate techniques to reduce the workload of the complex task of interpolation [130, 100, 121] are necessary.

2.2.2   Persistent Storage and Archiving

Besides the requirement of (near) real-time monitoring of phenomena, the acquired data, or at least parts of it, have to be stored permanently for subsequent analyses and considerations of long-term trends. For real-time monitoring it is essential to provide a model that is as accurate and up-to-date as possible under given circumstances and restrictions. The requirements for long-term storage are even more challenging since a good compromise has to be reached between the following partly conflicting requirements:

- sufficient quality and density
- small data volume
- originality of the data
- consistency
- informative metadata
- quick and intuitive data retrieval

To be beneficial for as many applications as possible, the data should be stored in spatio-temporal databases with unambiguous spatio-temporal reference systems.

Which data should be stored depends very much on the type of application that is planned for retrieval. Generally, it is advantageous to keep as much of the original sensor data as possible (originality). So if knowledge that is unavailable at the moment of observation—like a large sensor drift—becomes available after archiving, it can be considered in forthcoming analyses. This is hardly possible if only the *derived data* like raster files are stored.

On the other hand, storing original sensor data means additional effort in the moment of query to provide it in a form that is suited for interpretation or analysis. So if a (n-dimensional) grid of the observed phenomenon is required, the original sensor observations will have to be interpolated according to that grid's resolution. Depending on the grid size and resolution and the available computing power, this might lead to significant delay for retrieval.

When efficient retrieval of processed data is of high priority, it might be the best choice to permanently store the data (redundantly) in grid format. There are mature techniques to organize the data management this way. Lossless or lossy compression can help to reduce necessary storage space [49].

Nevertheless, there is a dilemma according to the management of the monitoring data. Whether to store the original vector data or derived raster data or both has consequences on volume, flexibility, usability, redundancy, consistency, responsiveness, to name a few. Just as with the process of monitoring as a whole, an appropriate solution has to be a compromise of multiple objectives in accordance with the goals. The concept of a field data type [76] is an important milestone on the way to a consistent storage schema for data about continuous phenomena.

2.2.3   Retrieval

For the retrieval of data, there are different scenarios that are reasonable in the context of environmental monitoring. As already mentioned, the monitoring system has the role of the mediator between the available observations and the required knowledge. It fills the gaps in space and time that necessarily remain between the available discrete observations. The overall quality of the model depends on the density of observations and the interpolation method. From this derived model, data retrieval can be categorized in different modes:

1. **Interpolated points:** Values of the variable of interest can be queried at arbitrary positions in space and time. This mode is applied when the variable is needed to examine some critical event.

   For example, when investigating an increased rate of short circuits of a particular model series of power inverters, the actual precipitation at the time and position of each incident might reveal an important hint. Beside the value itself, the estimation variance—as provided by kriging—might also be important to judge the situation.

2. **Interpolated grids:** For visualization or intersection with other data, some representation of the variable as equidistant array of values is needed. For example, a time series of values for every quarter of an hour could be

derived from irregular observations to match the schedule of some other variable to investigate correlations (e.g. air pollution and rainfall). A two-dimensional grid of interpolated values provides a map of the phenomenon at a particular moment in time to be interpreted geographically. Adding the temporal dimension would produce a simulation of the phenomenon as a *movie* [131], as known from weather forecasts.

3. **Aggregations:** In order to get some summarized information of a region of particular extent in space, time or space-time, the interpolated grid can be aggregated to the required value. The monthly average value of a pollutant in a particular district is one example of such an aggregation. It presumes an intersection of the interpolated grids (see above) with the (spatio-temporal) target feature. Other indicators like minimum, maximum, median or variance might also represent valuable information. Such aggregations are crucial as an integral element of an alert mechanism within the monitoring system.

As these modes above show, the retrieval of data about continuous phenomena has specific characteristics which are not necessarily covered by standard GIS functionality. What is needed for interoperability of systems in this context is a query language that abstracts from the format that the data is actually stored in. It has to provide formal expressions to describe spatio-temporal reference grids, their resolution and extent as well as definitions for aggregations, like monthly average values within a district. With such a query language [76], the client does not have to know about the format of the actual data but communicates with the system on a more abstract level (see Chapters 4, 6).

Additionally, some metadata will also have to be available in order to judge the appropriateness of the data for the given task. Value bounds, mean value and variance might be valuable informations for users of the data. On a more sophisticated level, the variogram can reveal the geostatistical properties of a dataset. Unambiguous identifiers and standardized formats are needed to retrieve and process these metadata.

2.3   Resources and Limitations

The description and prediction of phenomena is the central concern of science [102]. In order to be feasible in practice, only those parts of reality are considered that are relevant for a particular question or task [14, p. 6], [116, p. 9], [73, p. 4], [13, p. 17], [47, p. 91]. This deliberate reduction of complexity is basically what modelling and simulation is about [9]. According to the intention (e.g. knowledge, safety, ecological and economic benefit), models are designed to answer the crucial questions raised within the particular problem domain.

Monitoring can be seen as as the process that feeds such a model with empirical data in order to align it with reality. The necessary effort for monitoring depends on the requirements according to the coverage and accuracy of the model. As for any project, the fact of limited resources will put a considerable burden on it. So any monitoring strategy is actually a trade-off between necessary costs and achievable benefits, whereas neither cost nor benefit can always be expressed in monetary units. For the costs, there are following aspects to be considered:

- costs to obtain, install, operate, maintain and dispose of sensors
- costs for the infrastructure necessary to keep the monitoring in operation (communication, processing, archiving, provisioning)
- human resources (administration, maintenance, adaptation of new technologies, research, cognitive and emotional effort, ...)

On the other hand, there are the various benefits made available through the gained knowledge:

- economic benefits when better predictions result in a more efficient process of generating added value (like e.g. for fishery, agriculture, forestry...)
- improved knowledge about the environment as the livelihood for present and future generations
- governmental healthcare (e.g. air pollution)
- better disaster management (distribution of toxic fumes, radioactivity)
- improved quality of life through information (forecasts for weather, pollen drift, ...)

In this context, science and technology can only try to help to explain phenomena, explore causalities, propose solutions, support their implementation and monitor their effects. To provide the necessary resources for this task is the responsibility of society and politics [13, p. 29]. Whether the efforts go along with the proclaimed aims and values should be continuously examined carefully. Science itself has to be rigorous and consistent to withstand being abused by political or economic interests [127, p. 585 ff.], [62, p. 19]. Otherwise, it will deprive itself of its legitimation in the long run.

Given these circumstances, the operator or operating team of a monitoring system needs to carefully balance the input resources against the output benefit. In order to facilitate well-reasoned decisions here, two major investigations have to be carried out:

1. Learning about the process and its complexity
2. Identifying the questions intended to be answered by the monitoring

The fundamental objective of any monitoring system is to find and establish the link between those two domains in an effective and efficient way. A good compromise between invested resources and derived knowledge has to be found [13, p. 11]. In the following, the means and aspects of monitoring will be introduced and discussed according to their limitations.

2.3.1 Sensor Accuracy

The accuracy of each single sensor measurement in some way affects the overall accuracy of the monitoring system. It has to be in accordance with the requirements of the system and should be specified.

Accuracy is a function of precision and bias [12, p. 11 ff.], [89, p. 60 ff.]. Precision expresses the degree of scatter in the data around a constant value while bias is the deviation of this value from the true value. Precision is associated with random errors while bias is caused by systematic errors.

In the context of monitoring systems there can be mechanisms to detect systematic sensor errors if the data is sufficiently redundant. In such a case, the sensor can be calibrated in order to produce correct measurements. If such a sensor error can be determined to first appear at a particular point in time, all registered measurements since then can be corrected retrospectively. To take into account such sensor errors for derived data like interpolated raster grids is much more expensive. This is a strong argument to favour field data types (see Chapters 2, 6, 8).

Yet, the incorporation of the sensor accuracy into the interpolation process has its own complexity and is not in the scope of this work. It is assumed that serious sensor errors are detected and considered by other system components. Small errors in the observations are generally assumed to be overridden by the inaccuracy caused by interpolation itself [98, p. 9].

2.3.2 Sampling

For the monitoring to be effective, the area of interest has to be covered by an appropriate set of observations. The sampling density and distribution must be sufficient to allow a reasonable interpolation at unobserved positions in space and time according to the monitoring objectives. It depends on diverse factors:

■ the phenomenon itself and its complexity (e.g. interdependencies with other variables)
■ the quality of the monitoring
■ the requirements of the monitoring (aggregation, reconstruction, archiving, retrieval, alert, ...)

The factors above are interdependent, which is shown by the following examples:

- The more complex a process is, the more observations are usually necessary to generate a model that adequately represents its dynamism
- The better the physical processes are understood, the less observations will eventually be needed to generate an appropriate model
- Changed requirements according to the aims of the monitoring will probably affect the overall effort that is necessary for the monitoring

The most crucial decision within a monitoring concept is about how the sampling is to be carried out. Insufficient sampling can not be compensated by even the most sophisticated interpolation method. So the region of interest should be covered by enough observations in order to capture the phenomenon accordingly.

On the other hand, within a monitoring scenario it might be the most expensive task to establish, operate and maintain the sensor network. So the other objective is to have as few sensors and observations as possible to achieve the required quality. To find a good compromise here is maybe the most important decision of the operator or operator team.

As already mentioned, there are basically two aspects to be decided for each sampling layout: the number of observations and their distribution. When assuming a regular pattern with a constant point distance to observe a particular region, the number of observations can be easily determined. A square grid is often applied, while a hexagonal grid is considered the more efficient variant [55, 22].

A regular pattern, however, can only rarely be applied because sensor sites are subject to geographical and infrastructural constraints (e.g. as for meteorological stations). The sampling pattern of mobile sensors like vehicles or drifting buoys changes continuously. Wherever static or dynamic sampling positions can be assigned freely, a deliberate selection of sampling positions (static) or active sampling (dynamic) should be considered [97, 10, 54, 126].

In this work, however, the sufficient sampling density is estimated for sampling positions that are randomly distributed in the region of interest. It is reproducible, easy to generate for any dimensionality and produces a big variety of sample distances, which is necessary for robust variogram fitting (see Section 5.3.5). A formula that derives this sampling density from geostatistical indicators is provided in Section 5.3.2.

2.3.3 Computational Power

Environmental monitoring of continuous phenomena faces limits of computational power according to the following tasks:

Data acquisition

Mobile units for processing and transmission have limited energy and therefore are also limited in their capabilities

Extensive processing

Massive observational data, complex interpolation algorithms and real-time requirements can raise the workload to a critical level

Due to cheaper sensor hardware and new sources like volunteered data acquisition, the availability of environmental observations is continuously increasing [56, 68]. Consequently, there is an ever growing computational workload for processing and analysis. State-of-the-art computer technology continuously provides more powerful and more energy-efficient machines. In recent years, increasing computational power is not achieved by higher clock speed but rather by increasing parallelization using multiple central processing units (CPU), graphics processing units (GPU) or field-programmable gate arrays (FPGA) [78]. This also affects software development since algorithms must apply multithreading techniques to exploit parallel processing architectures [23, p. 772 ff.].

So the challenge for acquisition of environmental data is to use the limited computational resources as efficiently as possible. There are usually several degrees of freedom on how to perform monitoring since the associated tasks of acquisition, processing and transmission can be carried out in different ways [42].

For complex calculations that are performed on powerful workstations, servers or even computer clusters, the focus lies on optimizing algorithms, data structures and indexing to achieve sufficient performance for processing and retrieval.

For both scenarios—the data acquisition and the complex processing— simulation can be used to test and evaluate several variants according to their overall computational efficiency. Therefore, the simulation needs to keep track of the computational work for each scenario (see Section 7.5). This indicator, among others, can be used for iterative optimization according to the prioritized goals.

2.3.4 Time (Processing and Transmission)

For many tasks in environmental monitoring, time is a scarce resource. There are complex analyses that have to be carried out in time in order to be valuable (e.g. pollution alert systems). Sensor observations need to be transmitted and analysed immediately to detect dangerous states and to limit damage [7]. More powerful hardware is one way to address this challenge, but it is not always feasible. Inappropriate costs or limited energy for wireless devices might be arguments against this option.

Nevertheless, the time factor needs to be considered carefully in such situations. So there is good reason to be able to keep track of it explicitly. For a given task, the processing time is basically determined by the workload of necessary CPU cycles, the CPU clock speed and the number of available processors. Compiler optimization, operating system and features of the programming language will also affect performance and should be considered where necessary.

A generic description of temporal effort therefore has to consider two major components: (1) a quantification of the workload of a task by the number of instruction cycles it entails and (2) a formal specification of the performance-relevant properties of the machine the task is processed on. Leaving aside parallelization, time effort for a particular task is basically determined by the ratio between the cycles of calculation and the CPU clock speed. In most cases however, the proportion of parallelizable code segments and the number of processors of the machine have to be considered as interdependent factors in order to be realistic (see Section 5.5).

For transmission, the data rate and the amount of data to be transmitted plus communication overhead will determine the necessary time. Compression and progressive retrieval [82] can reduce the transmission time and consequently the energy expense (see Section 5.4.3).

2.3.5 Energy (Processing and Transmission)

Particularly when tasks like observation, transmission or processing are performed on battery-operated devices, the energy consumption of a monitoring scenario has to be considered to achieve an efficient use of resources [66]. In a distributed scenario, there are usually several degrees of freedom to fulfil a monitoring task with respect to how and where to perform the several operational steps.

A simple example for that is the exchange of data in a sensor network. One option is to transmit the original data without further processing, the other is to compress the data before transmission and decompress it after receiving it. There is significant computational effort for the compression and decompression process, but since wireless data transmission is much more energy intensive, this method will usually pay off [6].

When energy consumption is to be estimated for a particular task, the number of instruction cycles to be processed is the key indicator, just as it is for time consumption when assuming a particular frequency. Furthermore, while the CPU clock speed is taken as a denominator when estimating time consumption, we need a factor that quantifies the energy consumption per instruction cycle here (see Section 5.5).

Similar to the estimation of time, the energy expense can be calculated for a given constellation of instructions and hardware specifications. Aspects like parallelization and idle mode energy consumption can make such estimations more complex.

The problem of energy efficiency in wireless sensor networks has been discussed extensively in literature [42, 66, 67, 121, 63]. When energy consumption is modelled within a simulation framework as described above, it can be determined for different monitoring strategies when processing them in simulations. This can be of vital importance for wireless constellations.

Data transmission is a critical issue for wireless sensor networks since it usually consumes much more energy than acquisition and processing of the data [42, p. 79], [66]. As already mentioned, wherever there are multiple feasible scenarios of how to collect, transmit and process the data for a monitoring, energy efficient variants should be chosen particularly for wireless constellations.

The energy demand for data transmission will depend on data volume, hardware, protocol, medium and geometrical constellation of the network. There is also potential for improvement of efficiency by adaptive configuration of the transmission process [77].

Such optimization should be carried out beforehand with the help of a simulation, which presumes that the aspect of transmission is adequately modelled according to the issues to be considered like geometric constellation or transmission schedules. A closer examination of this problem is not in the scope of this work.

2.4 Summary

The limitations discussed above are challenging when establishing a system for environmental monitoring. The responsible actors or decision makers need to identify and precisely formulate goals and priorities and to deliberately choose the appropriate devices, methods and configurations to fulfil them. In order to support this complex task, it is helpful to begin with structuring the problem on an abstract level [47]. The very general properties involved when establishing an environmental monitoring system are: the aims to achieve, the required quality and the generated costs, as illustrated in Figure 2.2.

All these components need to be considered carefully to establish a monitoring system that fulfils the given requirements. Changes in one component usually will affect the other ones, which is indicated by the connecting lines.

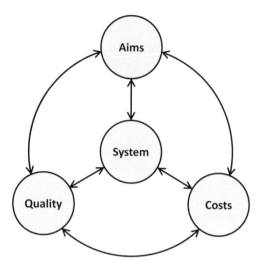

Figure 2.2: Superordinate monitoring system properties and their interdependencies.

Each of these superordinate properties entails issues that may or may not play a role for the particular monitoring task. Some of them are listed below for each property:

Aims:

- economic benefit
- scientific progress
- foundations for better planning
- political arguments
- environmental protection
- security and healthcare

Quality:

- coverage
- accuracy
- resolution
- availability
- interpretability
- interoperability (standard conformity)
- response time

Costs:

- infrastructure
- finances
- time
- organizational effort
- cognitive/mental effort
- environmental impact (e.g. stations and their maintenance)

System:

- sensors
- processing units
- (wireless) network
- protocols
- formats
- standards
- methods/algorithms
- parameters
- performance indicators

To bear in mind all of the relevant aspects from the listing above is already challenging. The interdependencies that exist between these aspects massively increases the complexity in a way that solutions usually can only be found iteratively in an evolutionary learning process [47, p. 64], [116].

In this sense, the intention of this work is to support such iteration concerning the aspects of sampling distributions and density, interpolation algorithms and associated parameters. Depending on the phenomenon and the requirements of the monitoring system, different constellations can be simulated and evaluated by output indicators that express both quality and costs.

The evolutionary process of acquiring new knowledge and improving the model is supported by the circular design or "closed loop" [116, p. 9] of the framework. By using synthetic data as reference, the *root mean square error* (RMSE) quantifies the fidelity of the derived model and thus is the crucial indicator of the monitoring quality [51, p. 114 ff.].

Other indicators like computational effort (see Section 5.5.2) are also used. Variations of methods and parameters can easily be processed in batch mode using the concept introduced in Section 5.5. Given the framework that is described here, it is easy to automatically perform multiple variations of system configurations. The application of this general concept will be outlined in this work after the method of spatio-temporal interpolation is introduced in the next chapter.

Chapter 3

Spatio-Temporal Interpolation: Kriging

CONTENTS

3.1	Method Overview	38
3.2	The Experimental Variogram	39
3.3	The Theoretical Variogram and the Covariance Function	40
3.4	Variants and Parameters	45
3.5	Kriging Variance	48
3.6	Summary	50

3.1 Method Overview

The general properties of the geostatistical method of kriging have already been introduced in Section 1.5. It assumes a stationary process (in practice however, only second order and intrinsic stationarity are relevant [94, 31]) and interpolates between observations by estimating optimal weights for them while taking into account their correlation according to their distances. This fundamental relation between the distance and the degree of correlation of two positions is expressed by the covariance function.

When it is applied to an actual set of observations, the method is fundamentally a two-step process [125]:

1. Inspection and mathematical description of the spatial, temporal or spatio-temporal autocorrelation of a given set of observations
2. Interpolation between the observations with respect to the detected autocorrelation structure

One might also say in other words: After specifying the rules of autocorrelation from the given observations, these rules are applied to estimate the value between the observations. Unlike the deterministic methods listed in Section 1.4, it incorporates the statistical properties of pairs of observations with respect to their spatial, temporal or spatio-temporal relation.

The inspection of the statistical properties is also carried out in two steps: (1) the generation of the experimental variogram and (2) the fitting of the theoretical variogram. These procedures will be explained in the next two sections.

The autocorrelation structure of a set of observations can be expressed abstractly by the variogram model and the associated covariance function, which has to be fitted to the empirical data. The value prediction for a given point is then performed by estimating the optimal weight for each observation while considering this autocorrelation structure [95, 8].

For simple kriging, the vector of weights λ is determined by:

$$\lambda = C^{-1} \cdot c, \tag{3.1}$$

where C is the quadratic covariance matrix generated by applying the covariance function to each observation pair's distance and c is generated by applying the function to the distances between the interpolation point and the observations, respectively.

The preceding step of fitting of an appropriate covariance function that is needed to populate matrix C and vector c will be outlined in the following.

3.2 The Experimental Variogram

Given a set of observations—often irregularly distributed in space and time—the primary aim of geostatistics is to inspect and describe its statistical properties in order to perform optimal interpolation.

To describe the autocorrelation of a given set of observations, the spatial, temporal or spatio-temporal distances for all possible pairs of observations are related to their *semivariances*

$$\gamma = \frac{1}{2}(z_1 - z_2)^2. \tag{3.2}$$

Given n observations, the number of pairs p is given by $p = (n^2 - n)/2$. For visual interpretation, for each pair of observations a point can be plotted in a coordinate system that relates spatial (ds) and/or temporal (dt) distances to the respective semivariances γ (see Figure 3.1).

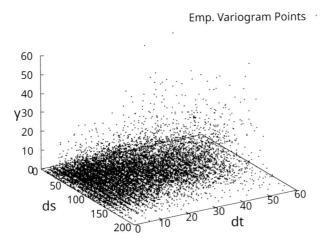

Figure 3.1: Spatio-temporal experimental variogram relating pairwise spatial distances (ds) and temporal distances (dt) to semivariance (γ).

The plot clearly reveals the fundamental characteristic of stationary phenomena: observations proximate in space and time tend to be similar in value while distant ones tend to scatter more.

The dimensionality of the variogram (both the experimental and the theoretical one) depends on the number of dimensions which are related to the calculation of the semivariance γ. Taking into account only the spatial distance—be it in one-dimensional (transects, time series) or two-dimensional space—leads to a two-dimensional variogram. The anisotropy—the dependence of correlation not only on the *distance* but also on the *direction*—can be considered by multiple variograms for different circular sectors or, for more precision, a three-dimensional surface [94, p. 100]. Anisotropy is not considered in this work in order to limit the overall complexity. It might easily be incorporated in both the random field generator and the variogram fitting.

Considering *time* also adds a dimension to the variogram as depicted in Figure 3.1. The autocorrelation structure can be interpreted here visually with respect to spatial and temporal distances. The differing characteristics of correlation decay in space and time and also spatio-temporal interdependencies [31, 53] can thus be inspected. A noticeable scatter near the coordinate origin indicates the *nugget effect* [129, 94], which is also not considered here for the sake of complexity. Its inclusion in the simulation framework would be carried out analogously to anisotropy.

To be applicable for calculation, the autocorrelation structure that is materialized in the experimental variogram needs to be expressed abstractly as a mathematical function. It is called the *theoretical variogram* and will be discussed in the next section.

3.3 The Theoretical Variogram and the Covariance Function

The theoretical variogram can be seen as mediator between the experimental variogram derived from the observational data and the covariance function needed for the population of the covariance matrices. There is a symmetric relationship between the theoretical variogram and the covariance function, as Webster & Oliver state in [129, p. 55]:

> Thus, a graph of the variogram is simply a mirror image of the covariance function about a line or plane parallel to the abscissa.

This relationship is apparent when comparing Figures 3.2 and 3.3.

The fundamental geostatistical parameters *sill* (which expresses the dispersion of values for distant points) and *range* (which expresses the distance up to which spatial autocorrelation takes effect) do exist for both representations. Therefore, the fitting of the theoretical variogram to the experimental variogram pointcloud also provides these parameters for the covariance function.

In this context, it is appropriate to point out the fundamental relationship between variance, covariance and correlation, as specified by Abrahamsen [1, p. 9]:

$$c(\tau) = \sigma^2 \rho(\tau), \tag{3.3}$$

where c is the covariance, σ^2 is the variance and ρ is the (normalized) correlation, given the separation vector τ as a variable.

The geostatistical parameter *sill* is sometimes falsely associated with the *dispersion variance*. For a stationary process, the *dispersion variance* is slightly less

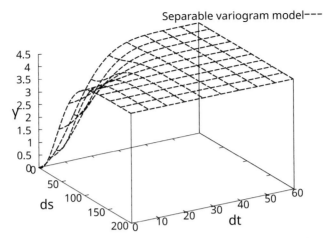

Figure 3.2: Spatio-temporal theoretical variogram relating pairwise spatial distances (ds) and temporal distances (dt) to semivariances (γ).

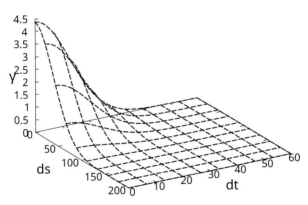

Figure 3.3: Spatio-temporal covariance function relating pairwise spatial distances (ds) and temporal distances (dt) to semivariance (γ).

than the *sill variance* [129]. However, since it is a parameter to be determined by iterative fitting (see Section 5.3.5), the dispersion variance can very well serve as a prior estimation for the *sill* variance.

Being an abstract model of a (spatio-temporal) dispersion structure (Figure 3.2), the theoretical variogram can visually be associated with the experimental variogram (Figure 3.1).

As the mirror function, the covariance function relates the covariance (and thereby also the correlation, see Equation 3.3) to the distance in space and time.

As can be seen in Figure 3.3, it has its maximum value at the origin and decays with increasing distance.

It depends on the characteristics of the phenomenon which variogram model (and therefore which associated covariance function) to choose: How many dimensions (spatial, temporal or spatio-temporal) are there to be considered? By which law does correlation decrease with increasing spatial and/or temporal distance? Is there any noise for distances near zero (nugget effect)? Does the correlation not only depend on distance but also on direction (anisotropy)? How are the spatial and the temporal dimensions entangled [31, 129]?

The basic two-dimensional representations of three commonly used covariance functions are depicted in Figure 3.4. Their behaviour especially near the origin and beyond the *range* point implies different characteristics of the corresponding process [129, p. 80 ff.]. An initially small (Gaussian) or moderate (spherical) slope that moderately increases (in absolute value) before decreasing again (Gaussian) represents a rather smooth model whereas a steep slope at the origin (exponential) indicates greater dynamics at a small scale. This can be comprehended from the graphs and their associated random fields in Figure 3.4. In contrast to the other two covariance functions, with the spherical function the correlation ceases to zero for distances greater than *range*.

The respective equations that generate the graphs in Figure 3.4 are given below with c = covariance, h = distance, s = sill and r = range:

Gaussian:

$$c(h) = s \cdot e^{-\frac{h^2}{\frac{r}{\sqrt{3}}^2}} \tag{3.4}$$

Spherical:

$$c(h) = s \cdot \left(1 - \left(\frac{3}{2}\frac{h}{r} - \frac{1}{2}\left(\frac{h}{r}\right)^3\right)\right) \ for \ h < r, \ else \ c(h) = 0 \tag{3.5}$$

Exponential:

$$c(h) = s \cdot e^{-3\frac{h}{r}} \tag{3.6}$$

A graph intersecting the ordinate below the *sill* value (1.0 in Figure 3.4) would represent a noise for near zero distances as *nugget variance* or *nugget effect*. Its representation in the *theoretical variogram*—being the mirror image of the covariance function—is more intuitive since the function starts with the value of the nugget variance on the ordinate.

For spatio-temporal kriging, the principle of correlation decay as explained above needs to be extended for two input variables, namely spatial and temporal distances. In the context of the simulation framework as described here, this has to be considered for three basic procedures: (1) the variogram-based filter that generates continuous random fields (Section 5.3.1), (2) the fitting of the variogram (Section 5.3.5), and (3) the kriging interpolation (Section 5.3.6).

When more than one dimension is used as explanatory variables of the variogram model, the interaction between the dimensions according to correlation

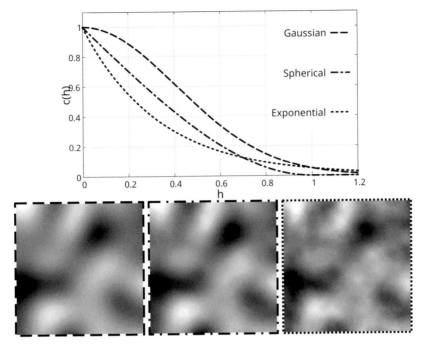

Figure 3.4: Different covariance function types (top) and their respective random fields (bottom) with corresponding dashed frames; as can be seen, the smoothness of each field is a function of the slope near the coordinate origin.

has to be specified (see [31, p. 297 ff.] and [53] for thorough study and derivation of the formulae below). Within the framework, four spatio-temporal variogram models commonly mentioned in literature have been implemented (see Figure 3.5).

In the case of *separable* covariance functions (see Figure 3.5(i)), the product of two *separate* covariance functions yields covariance values for compound distances in space *and* time with

$$C(s,t) = \sigma^2 f(s)g(t), s \in \mathbb{R}^2, t \in \mathbb{R}, \tag{3.7}$$

where σ^2 is the sill variance, $f(s)$ is the covariance function for the spatial component and $g(t)$ is the one for the temporal component. They might differ in mathematical models and associated parameters to reflect different dynamics in space and time.

In *nonseparable* variants (Figure 3.5(ii)), the spatial and temporal components are entangled by

$$C(s,t) = \sigma^2 exp\{-k_s^2 \|s\|^2/(k_t^2 t^2 + 1)\}/(k_t^2 t^2 + 1)^{d/2}, s \in \mathbb{R}^2, t \in \mathbb{R}, \tag{3.8}$$

where k_s and k_t represent scaling parameters for the spatial and temporal component, respectively, and d stands for the number of spatial dimensions [31, p.

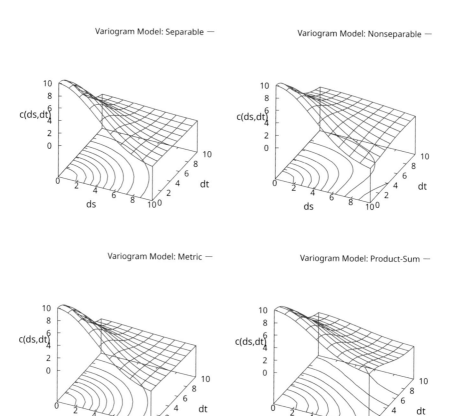

Figure 3.5: Spatio-temporal variogram models: (i) separable, (ii) nonseparable, (iii) metric and (iv) product-sum, plotted as covariance functions with parameters spatial distance (ds) and temporal distance (dt).

317]. The term reflects spatio-temporal interaction that can actually be found in many physical processes [31, p. 309 f.].

The *metric* covariance function (Figure 3.5(iii)) simply applies a spatio-temporal anisotropy factor (k_t) to align the temporal with the spatial dimension:

$$C(s,t) = \sigma^2 f(\sqrt{\|\mathbf{s}\|^2 + (k_t|t|)^2}), s \in \mathbb{R}^2, t \in \mathbb{R} \qquad (3.9)$$

In contrast to that, the product-sum model introduces some more interaction between space and time (Figure 3.5(iv)):

$$C(s,t) = k_1 f(s)g(t) + k_2 f(s) + k_3 f(t), s \in \mathbb{R}^2, t \in \mathbb{R} \qquad (3.10)$$

The equations above express the autocorrelation structure of a random field according to space and time. They are used in this work for generating random fields, choosing and fitting variogram models and interpolation.

In the experimental setup proposed here, the random fields are generated by a filter kernel that applies a separable variogram model. A systematic and thorough investigation of dependencies between the applied variogram *filter*, the applied variogram model and its parameters and the resulting accuracy (RMSE) might reveal interesting dependencies, but is out of the scope of this work. There is, however, a relation between the range value used for the variogram filter and the one determined by the variogram fitting procedure and subsequent kriging that is applied to the random field generated by it: The closer the estimated *range* value is to the one of the variogram filter, the better the interpolation results become (see Section 7.2).

3.4 Variants and Parameters

Kriging has evolved to a complex technique with an almost overwhelming amount of varieties and associated control parameters. Due to this complexity it is often difficult to decide whether it is applied and configured correctly; the mere selection as a method does not sufficiently imply appropriateness or inappropriateness [89, p. 42 ff.].

An overview of the most used versions is given by the list below without any claim of completeness. It is mainly based on the rewiev of Li & Heap [75]; see also [75, 19, 129, 31] for further study.

Block Kriging

In contrast to point-oriented estimations, block kriging (BK) claims interpolations for (n-dimensional) regions of arbitrary form.

Cokriging

Cokriging is the multivariate version of kriging that exploits cross-correlations between different variables (e.g. atmospheric pressure and precipitation) to improve predictions.

Disjunctive Kriging

Disjunctive kriging transforms the primary variable to polynomials that are kriged separately and summed afterwards. It is applied when the primary variable does not sufficiently represent a normal distribution.

Dual Kriging

Instead of the values themselves, this variant estimates the covariances. It is used when the filtering aspect of kriging is of interest.

Factorial Kriging

By applying nested variograms, factorial kriging can combine different correlation structures at different scales.

Fixed Rank Kriging

This variant is applied for big datasets and reduces the computational workload for inversion of the covariance matrix.

Indicator Kriging

When the output variable is supposed to be binary, representing some threshold (e.g. humid vs. arid), indicator kriging can be applied.

Ordinary Kriging

Ordinary kriging incorporates the estimation of the mean value by adding lagrange multipliers to the covariance matrix.

Principal Component Kriging

Principal component analysis (PCA) is used to identify and quantify correlations in the data, then process the identified (uncorrelated) components separately, before generating the estimation by linear combination of these components.

Regression Kriging

In regression kriging, any trend is determined and removed from the data before the interpolation and added again afterwards.

Simple Kriging

Simple variant of kriging that presumes a constant and known mean value. Given the synthetic random fields generated according to this and other statistical parameters, this is the variant that was predominantly used in this work.

Universal Kriging

To integrate a trend in the process, universal kriging incorporates a smooth surface as a function of position.

The variants listed above are by no means exhaustive but can only give a hint of the versatility of kriging, which emerges from combinations and subclasses. Unlike some more intuitive geographic analysis tools (e.g. intersection, buffering, spatial join among others), kriging as a method requires deeper understanding of the underlying principles to be applied appropriately [89]. This is also the case for its control parameters, of which the most important ones are listed below.

Sill

As already mentioned, the *sill variance* or *sill* expresses the overall variability of a random field. It represents the maximum threshold value for semivariances of pairs of observations (Equation 3.2) and is generally reached at distances exceeding the *range* parameter (see below). The *sill variance* is not to be confused with the *dispersion variance* which is just the variance of observations in the classical sense. The actual *sill variance* should rather be estimated by fitting the theoretical variogram to the data. However, since this procedure is usually performed by optimization techniques (see Section 5.3.5), the dispersion variance can be a good first approximation.

Range

The *range* is the second decisive parameter for kriging since it expresses the distance up to which the observations are correlated stronger than the average of all possible pairs of observations [129, p. 89]. In the case of spatio-temporal variogram models, there might be separate *range* parameters for the spatial and temporal dimensions. For the separable variogram there is a *range* parameter for the spatial and one for the temporal covariance function (Equation 3.7). In the case of the metric variogram model, the spatial and temporal distances are cumulated by using an anisotropy factor (Equation 3.9). For the nonseparable variogram (Equation 3.8) and the product-sum variogram (Equation 3.10) there are factors to control scaling *and* interaction of space and time [31].

Nugget Effect

Just as for the range, the nugget effect also might show different dynamics in space and time, resulting in a joint short-distance noise. The nugget effect, however, is often difficult to estimate in practice since the measuring stations are chosen at a distance to avoid redundancy for economic reasons [53]. As mentioned before, the nugget effect is not considered further in this work.

The parameters above are usually estimated specifically for each dataset by fitting the theoretical variogram to the respective experimental variogram. The more parameters take part as variables in this fitting procedure, the more cumbersome the optimization can become. This aspect will be addressed in Section 5.3.5 and Section 5.5.

Given the variety of methods and parameters mentioned above, it is worth considering an architecture that provides the interpolation of a value of interest as a service. Without having to deal with too many details and program specifics, a common method with approved configuration could simply be identified by a unique name. Alternatively, for more flexibility the service could be configured by an appropriate interface (see Chapter 8).

3.5 Kriging Variance

Apart from the advantages of the method of kriging that have been covered so far, the provision of the estimation variance is unique among interpolation techniques [95, p. 1, p. 60]. It is a by-product of each point interpolation and reflects the uncertainty of estimation resulting from the constellation according to spatial and temporal distances to observations. Just as the estimated variable itself, it represents a continuous field that can be discretised as raster grid.

The kriging variance is derived by multiplication of the weight vector from Equation 3.1 with the vector c that contains the covariance values of each observation with respect to the interpolation point:

$$v = \lambda^{\mathsf{T}} c \qquad (3.11)$$

As the ingredients of the derived value show, it only depends on the covariance structure that is given by the geometric constellation of the observations and the interpolation point; it does not depend on the observed values themselves [54].

When creating raster grids by interpolation with kriging, it is useful for many purposes to also store the kriging variance (or deviation) for each cell as an additional dimension or channel. This "map of the second kind" [89] reflects the confidence of estimations as a continuum with respect to the proximity to observations.

In the scope of this work, the kriging variance represents highly valuable information. In the context of monitoring it—or its complementary value—can also be interpreted as *information density* and thus be exploited to address several problems of monitoring listed below:

Continuous integration of new observations

In monitoring scenarios where a state model has to be provided in (near) real time, there is the problem of how to seamlessly integrate new incoming observations. For workload and consistency reasons, this updating should be carried out without having to (1) calculate the model anew using all previous *and* new observations or (2) replace the old model by relying only on the most recent but probably very few observations. A compromise would be a sliding window [131] containing only observations that do not expire. However, depending on the spatio-temporal distribution of the observations and the size of the time window, the approach might cause temporal discontinuities if the window is too small and heavy computational workload will be required if it is too large.

Alternatively, the model can be updated smoothly and selectively wherever new observations occur. To accomplish this, the kriging variance—continuously available for the previous and the new model—can be used as weighting schema by which both models are merged (see Section 5.4.2). This method is highly flexible in terms of the number of new observations to be integrated because it retains the previous model where no new information is given instead of indifferently overwriting it.

Performance improvement by model subdivision

Apart from its application for continuous updating in a data stream environment as described above, the method can also be used to mitigate the computational burden of numerous observations. Instead of including all observations in one large model, it can be divided into subsets that are processed individually and then merged by weights based on their kriging variances (see Section 5.4.2, also [81]).

Confidence about critical state checks

In a monitoring scenario with critical state checking (see Figure 5.13), the kriging variance can significantly help to put such a statement on an objective basis. So if a sensor network is installed to push an alert in case of some exceeded threshold, the *intrinsic* idea behind it is to permanently exclude the possibility of that threat. Whether this is the case because of an actually exceeded threshold or because some sensors are down and therefore no sufficiently secure knowledge is available: some actor needs to be notified to induce some predefined procedure. The failure of sensors might eventually not change the derived *value* itself, but rather its *variance* and therefore the confidence of the associated state check (see Chapter 6, Section 7.6).

Adaptive filtering

Data sparsity is a very common problem for monitoring scenarios. However, with an increasing number of available low-cost sensors just the opposite can become a problem that calls for *decimation* of observations. It should be carried out deliberately since autonomous mobile sensors might not be distributed homogeneously (like drifting buoys, see [130]). Observations that would only minimally contribute to updating the model should preferably be left out. The kriging variance map as indicator for *information determination* or *information density* can be used for such adaptive filtering: only observations above a particular variance threshold are considered in order to limit data redundancy. Another way would be to order a set of new observations by the values determined by their position on the kriging variance map to leave out a particular number of observations or portion of the data. The utilisation of the kriging variance as filter provides a flexible and adaptive solution wherever too much observational data is a problem.

The various areas of beneficial application of the kriging variance or *kriging error maps* [89] as listed above constitute a strong argument in favour of kriging as an interpolation method. The estimated confidence for each interpolated value is such a crucial information that it should always be considered carefully.

3.6   Summary

Notwithstanding the computational burden kriging lays on the monitoring system, it offers several features that make it unique compared to other interpolation methods:

- It is an unbiased estimator of minimum variance [95]
- It is not only well established in geosciences, but also in the area of machine learning, where it is known as Gaussian process regression [106, 46]
- By the concept of the variogram and the associated covariance function, kriging allows the consideration of even complex correlation structures with respect to time, space, space-time, periodicity, nugget variance, anisotropy among others [129, 31]. Given this powerful feature, the method is capable of adapting to a large variety of phenomena
- There is a vast number of kriging variants to address the wide range of problems associated with the monitoring of continuous phenomena [75, 19], [89, p. 43]
- The parameters of the variogram are usually estimated from empirical data; they specify the statistical properties of a particular phenomenon; their values might provide valuable information for retrieval when provided as metadata
- The kriging variance with the associated kriging error map or the "map of the second kind" [89, p. 464] is crucial where confidence of the interpolated values is important. It can also be exploited for features like performance improvement, continuous seamless updating or filtering

The undisputable high computational burden ($\mathcal{O}(n^3)$ for the inversion of the covariance matrix) of kriging may disqualify the method where high throughput goes along with real-time requirements. In such cases, inverse distance weighting (IDW) might be preferable due to its lower complexity [131].

On the other hand, its often superior interpolation quality [6, p. 51], the explicitly estimated and intuitively interpretable statistical parameters, and the very valuable additional information of the kriging variance make it a choice that should always be considered carefully. For operating an environmental monitoring system it provides sophisticated means to address many problems that occur in this context. With its diverse variations and parameters it is well suited for iterative improvement within a simulation environment.

Chapter 4

Representation of Continuous Phenomena: Vector and Raster Data

CONTENTS

4.1   Overview 52
4.2   Vector Data Properties 55
4.3   Raster Data Properties 56
4.4   Raster-Vector Interoperability 57
4.5   Summary 60

4.1 Overview

In the history of GIScience [50], there have been several different concepts to capture and represent phenomena of the real world. Apart from imaging methods like remote sensing, which are not covered here, data about continuous phenomena are increasingly acquired by stationary sensor observations [35, p. 66]. To be put in context, these observations have to be placed in some spatio-temporal reference system. They quantify some particular property or properties for that position.

Such discrete observations, however, mostly do not represent the continuous phenomenon of interest in an appropriate way. As originally applied in computer graphics, the colour of each pixel of an array represents some attribute, often considered continuous, like elevation or temperature [41, p. 75]. Such representation will broaden the applicability of the given observational data and thus increase their value. This is the case for visualization, but also when this representation is in some way to be overlaid with other data (like topography, demography, traffic and others). Such overlay or intersection is crucial for the generation of new knowledge within a geographic information system (GIS) [91].

In many aspects, the modelling of continuous fields shares diverse characteristics with other representations of environmental phenomena. Spatio-temporal referencing is the fundamental requirement for any system representing real-world objects. For objects which are considered rather discrete within a particular modelling context, like a tree or a wall, the main problem is to determine their position in some coordinate system that is globally valid (geocoding [79, p. 103]). The change of structures over time—as caused by erosion—shall also be considered, but representation of strong dynamism is due to specific systems and is not typical for topographical mapping.

With continuous phenomena however, variability in space and time is crucial for two aspects:

1. The phenomenon itself is gradually changing when moving in space and time.
2. The phenomenon is observed at discrete positions in space and time. These observation positions typically do not sufficiently cover the positions and regions of interest as needed for analysis.

So the central challenge of a monitoring system in this context is to obtain some aggregated knowledge about a dynamic continuum while bridging space and time by some form of interpolation. The general principle of this task is illustrated in Figure 4.1.

Within the process chain of sampling, interpolation and aggregation, there are various degrees of freedom that determine both quality and costs of the monitoring process. So while the observations of the thermometer and rain gauge of the recent days might suffice to decide about hay harvesting, the differentiated dis-

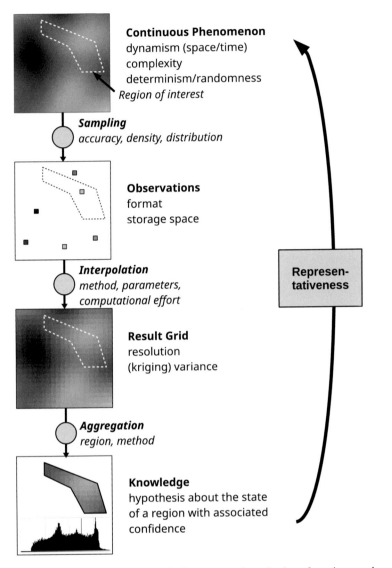

Figure 4.1: Raster/vector interoperability in the context of monitoring of continuous phenomena. Static states or datasets are labelled in bold, processes are labelled in italics. The phenomenon is captured by discrete observations that are interpolated and aggregated. This aggregation can then be used to accept or reject a hypothesis about the state of the phenomenon. The quality of the whole process determines its representativeness.

tribution of fertilizer (precision farming) presumes extensive and sophisticated data about previous crop sizes. In both cases the monitoring mediates between raw observations on the one side and decisions to be made based on derived knowledge on the other.

As can be seen in Figure 4.1, there are various applications based on the vector data format as well as on the raster data format in the context of the monitoring of continuous phenomena. These are listed below.

Vector format applications:

- Observations are usually considered as discrete events in space-time and therefore are related to some spatio-temporal reference system. In most cases, the observation of a particular variable is referred to a particular point within such a reference system.
- Subsequent observations from constant positions produce time series while multiple observations at the same time but different positions produce transects or grids.
- Observations from free floating objects along trajectories are dynamic according to their position in space *and* time.
- Spatial, temporal, or spatio-temporal regions that are subject to some investigation or aggregation are often confined by using polygons (space) and intervals (time). By intersection with the raster grid, the grid cells within this area can be identified and represent the subset from which an aggregation is to be calculated.

Raster format applications:

- The observation of the phenomenon might be carried out to a similar extent and resolution as for the final product, namely the image. One example for this is remote sensing. Such images need to be corrected geometrically by geocoding. This case is not considered further here.
- The result of an interpolation is often represented as a georeferenced grid. The resolution of the grid has to be a compromise between the desired granularity and the burden for processing and storage.
- The raster grid becomes the multi-dimensional bearer of the information that is originally provided by the observational vector data. Presuming a sufficient extent and resolution, it facilitates to combine this information with other datasets. Relevant subsets can be identified within the grid by selecting the affected cells through (spatio-temporal) intersection with the region of interest. Aggregations of various types can easily be derived from these subsets.

Depending on the task to be fulfilled with the help of a geographic information system, the representation of an aspect of reality by raster data might be more reasonable than by vector data or vice versa. For complex questions—like the one represented by Figure 4.1—this dichotomy [41] needs to be bridged in favour of interoperability. A functional integration of both approaches within one consistent query conceptualization might facilitate a generic standard for such complex state descriptions.

4.2   Vector Data Properties

As already mentioned in Section 4.1, the vector data model fulfils several tasks in the context of the monitoring of continuous phenomena:

■ Stationary sensor data are usually acquired, transmitted and archived as point data and related to some spatio-temporal reference system.

■ The area of interest is often represented as a polygon.

In contrast to raster data with its regular structure determined by resolution and orientation, vector data is much more suited for irregular distribution and flexible geometry definition as needed for the tasks above. The three geometric primitives *points*, *lines* and *polygons* [79, p. 162] will in this context represent single observations (points), transects and trajectories (lines) and areas of interest (polygons).

Vector data is very flexible with respect to the attributive data related to a geometric feature [79]. Sensor observations might be rather complex, so consequently the respective data structure should be capable of expressing such complexity. So there might not only be one but multiple variables to be acquired in order to provide rich information about the environment. The association between the observation and the particular sensor device is crucial when considering signal drift or other systematic errors for post-processing of the data.

The association between geometry and attributes, also known as georelational data, does allow for as many combinations of logical entities as is provided by the relational data model [17]. This powerful feature of the vector data model makes it indispensable also in the context of monitoring continuous phenomena.

So while the strength of raster data is to represent continuous phenomena in a contiguous way, the advantage of vector data is to represent and process particularities. It is suited to capture and preserve the raw form of information in a way that is not necessarily conditioned to be directly used for complex analyses. Derived formats and datasets like raster grids will serve this purpose. But in many cases those can be seen as functionally dependent from the original unprocessed observational data.

In a complex information system that is handling environmental data it is crucial to draw the distinction between original observational data and datasets that are derived from these by applying particular algorithms and associated parameters. The derived data should be treated as temporary because there might occur situations that make it necessary to generate them anew. Wrongly calibrated sensors, additional observations that were not available before, better interpolation algorithms or improved parameter estimations are situations that would justify such effort.

So the unprocessed original observations are the ones to be stored permanently. In the case of stationary sensors, these are vector data. In the case of continuous phenomena, the data have specific properties with respect to their distribution of values that can be exploited for data compression. This aspect is covered in Section 5.4.3.

4.3 Raster Data Properties

Due to its regular structure, the raster data format is suitable for representing continuous phenomena. If generated without *null* or *no data* grid cells, it immediately provides values at arbitrary positions within the region it covers. In the context of continuous phenomena, the generated raster grid can be seen as a kind of cache for the computational effort that is carried out for interpolation. It should be used as such wherever responsiveness is critical.

A more obvious application of raster grids as representation of continuous phenomena is their *visualisation*. If mapped to an appropriate colour ramp, the numerical values of the grid can be interpreted intuitively. Geographic information systems (GIS) do provide different colour ramps for this reason. When generating a raster dataset by interpolation, the resolution should be chosen carefully as a compromise between precision on the one hand and resource requirements for computation and storage on the other.

For remote sensing, raster files can contain several images or bands representing different colour channels. When used to represent continuous phenomena, there are aspects other than the observed value that can be stored in additional bands. For interpolation results generated by kriging, the kriging variance is maybe the most useful aspect to be stored alongside the value itself (see Section 3.5, Chapter 6). As a by-product of the interpolation process it is also continuous in nature and transports intuitively interpretable information about confidence—or information density—when rendered as image.

In the context of simulated monitoring as described in this work, the deviation of the interpolated model towards the reference model is a crucial indicator for evaluation of the interpolation and associated parameters (see Figure 5.3, p. 69). When plotted for each raster cell as image, it allows for visual identification of weak spots of the interpolation process as a whole.

Due to their regular distribution, the cells of a raster are suited as input data for representative regional aggregations. This is the case for all attributes mentioned above: the value itself, its kriging variance and its deviation from a reference model. An aggregation of values (e.g. their mean value) provides condensed information about that particular region as represented by the grid cells within that region. The aggregated variance expresses the overall reliability of that information. The root of the summed squared deviations of each interpolated cell with regard to the cell of the reference model—also known as root mean square error RMSE—serves as the overall quality indicator for a region.

The central property of a raster dataset is the coverage of a region with attribute values of homogeneous density. This density of values or resolution should meet this requirement to appropriately capture the dynamism of the observed phenomenon. An appropriate resolution is of course only a necessary but not a sufficient condition for an accurate model: only if there are enough observations and a valid interpolation method, the resulting grid will represent the phenomenon adequately.

The combination of a homogeneous and complete coverage of a region by values and their aggregation within a confined area do qualify raster grids as mediators between discrete observations and real knowledge. *Interpolation* disseminates observational information with respect to the statistical properties of the phenomenon, while *aggregation* relies on this homogeneously distributed data and provides a summation that represents the region appropriately (see Figure 4.1). In contrast to the aggregation of the observations themselves, this method takes into account the statistical distribution of the observed value as is expressed by the experimental variogram (see Section 3.2).

So while the vector format is the bearer of original observational data, raster grids hold the information derived from computationally expensive interpolation. In the case of kriging, the valuable information about the estimation variance can be provided alongside as an additional band.

Each format, vector and raster, has its own characteristics which have been laid out in this and the previous section. The requirements of sophisticated analyses can only be met if the capabilities of both formats are combined. Instead of regarding the associated operations as subsequent manual processing steps, there will probably be development towards an integrated and declarative query logic for continuous phenomena.

4.4 Raster-Vector Interoperability

In the context of continuous phenomena, vector data is used to represent discrete observations, transects, trajectories, and regions of interest; raster data disseminates such discrete information using regular grids of homogeneous density (see Section 4.1). Only a combination of these two formats can satisfy the requirements which are imposed on a sophisticated monitoring system.

In Figure 4.2, the interoperability of vector with raster datasets is sketched. In this schema, vector data contains the *dots of knowledge* within an observed area. Raster data fills this area with a regular pattern and thus distributes the knowledge in space and eventually time, thus making it available for further analysis like overlay with other data or for aggregation. The aggregation of the specific type is generated for a particular region, which is then again represented by vector data.

So in the scenario as described above, the raster data serves as an intermediate format capable of distributing and aggregating knowledge about a continuous phenomenon. It could be considered as a dataset to be stored temporarily only.

This aspect is illustrated by Figure 4.3, where the vector-raster-interoperability is shown from a perspective that is more operational: given the data observations and the region of interest as input and the interpolation parameters as control

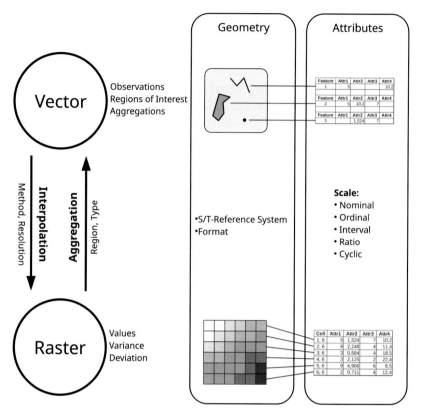

Figure 4.2: Vector-Raster Interoperability: properties and transformations are shown here in the context of continuous phenomena. Vector data are transformed to raster given the interpolation method and resolution. Raster data can be aggregated to represent a region of interest by a particular type like mean or variance. Raster and vector data share their dependency on a spatial and/or temporal reference system and a specific file or database format. The attributes are associated with pixels for raster datasets, and with geometries or composites of geometries for vector datasets. The observed property is expressed by these attributes using measurement scales.

variables, the process of interpolation and aggregation produces statistical indicators as output.

The geostatistical parameters sill and range are derived from the observations here, but they might also be subjoined from an archive (see Figure 7.18, page 161).

So within this scenario, the raster format provides the basic structure to distribute knowledge on the one hand, and to aggregate it again on the other. The way it does so should be reproducible by the parameters, so apart from computational efficiency there is no reason to keep these data permanently.

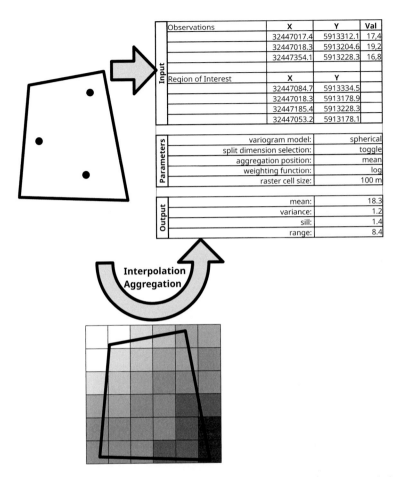

Observations	X	Y	Val
	32447017.4	5913312.1	17,4
	32447018.3	5913204.6	19,2
	32447354.1	5913228.3	16,8

Region of Interest	X	Y
	32447084.7	5913334.5
	32447018.3	5913178.9
	32447185.4	5913228.3
	32447053.2	5913178.1

variogram model:	spherical
split dimension selection:	toggle
aggregation position:	mean
weighting function:	log
raster cell size:	100 m

mean:	18.3
variance:	1.2
sill:	1.4
range:	8.4

Interpolation Aggregation

Figure 4.3: Monitoring Interoperability: While the vector data embodies the knowledge expressed by observations and the definition of the region of interest (*Input* part), the raster data is the destination format for interpolation and the source format for aggregation. The interpolation process is controlled by specific variables (*Parameters* part). The results relate to the region of interest and provide basic statistical indicators (*Output* part), of which *sill* and *range* are specific to geostatistics.

In the context of the monitoring of continuous phenomena, rasters can be considered as the (discrete) materialization of the concept of a field [25], [76], [41]. The resolution (represented by cell size in Figure 4.3) determines the degree of accuracy loss caused by approximation. Instead of stating this parameter arbitrarily, it might be derived from other indicators in the process.

The geostatistical parameter *range* might be a good candidate here, since it stands for the distance up to which there is a spatial correlation of values. So a small enough fraction of this distance might be a good estimate for raster cell size. It can be determined consistently for different sets of data while taking into account their actual dynamism.

Just as the control variable *cell size* mentioned above, other parameters also could be derived from the given data. So, for instance, the variogram model could be determined by cross-validation or other methods [89]. The downside of this approach is that it is computationally more expensive.

One way to overcome this problem might be to identify patterns in the data that indicate the use of particular parameters without having to perform expensive algorithms like cross-validation. Such patterns or hints should be expressed declaratively in order to be independent from particular software implementation and to make the decision rules explicit instead of hiding them deeply in the source code. The patterns in the data and parameter settings associated with them, respectively, provide a set of heuristics for adaptive data processing.

In an experimental setting as described Section 5.3 and applied in Chapter 7, such heuristics can be developed systematically over time. This evolutionary approach is designed to be independent from the applied interpolation method. It is based on a universal framework for parameter variation (for nominal, ordinal, interval, ratio or cyclic scales) and quality indicators produced by the interpolation process.

4.5 Summary

When monitoring continuous phenomena, the sensor observations are usually represented as vector data within a spatio-temporal reference system. In order to distribute the knowledge given by these observations, they are interpolated to a raster grid of sufficient resolution. From that raster, or a confined region within it, different aggregations can be generated in order to obtain some summarized knowledge. This knowledge is associated with the vector geometry representing the given region of interest.

By the interactional and transformative relationship between raster and vector data as described in this section, it is possible to express complex states and conditions of continuous phenomena. The automated selection of appropriate methods, parameters and grid resolutions is a challenging endeavour in this context. It has to be addressed if queries about states of continuous phenomena are to be formulated on a higher abstraction level without having to deal with details about interpolation parameters.

The overall workflow of a monitoring process following this objective is depicted in Figure 4.1. To obtain the required knowledge about a particular phenomenon, there are several conditions that have to be fulfilled. The amount and distribution of observations has to be sufficient to capture the phenomenon by the appropriate granularity. The interpolation method fills the gap between those observations by being applied to a regular grid. The relevant grid cells are finally aggregated to represent the knowledge as required.

Chapter 5

A Generic System Architecture for Monitoring Continuous Phenomena

CONTENTS

5.1 Overview 63
5.2 Workflow Abstraction Concept 64
    5.2.1 Datasets (Input/Source and Output/Sink) 66
    5.2.2 Process/Transmission 67
5.3 Monitoring Process Chain 68
    5.3.1 Random Field Generation by Variogram Filter 70
    5.3.2 Sampling and Sampling Density 73
    5.3.3 Experimental Variogram Generation 79
    5.3.4 Experimental Variogram Aggregation 80
    5.3.5 Variogram Fitting 85
    5.3.6 Kriging 88
    5.3.7 Error Assessment 88
5.4 Performance Improvements for Data Stream Management 89
    5.4.1 Problem Context 90
    5.4.2 Sequential Model Merging Approach 91
        5.4.2.1 Overview 91
        5.4.2.2 Related Work 92

	5.4.2.3	Requirements	92
	5.4.2.4	Principle	93
	5.4.2.5	Partitioning Large Models: Performance Considerations	95
5.4.3		Compression and Progressive Retrieval	98
	5.4.3.1	Overview	98
	5.4.3.2	Related Work	99
	5.4.3.3	Requirements	99
	5.4.3.4	Principle	100
	5.4.3.5	Binary Interval Subdivision	100
	5.4.3.6	Supported Data Types	101
	5.4.3.7	Compression Features	103
5.5		Generic Toolset for Variation and Evaluation of System Configurations	105
	5.5.1	Context and Abstraction	106
	5.5.2	Computational Workload	109
	5.5.3	Systematic Variation of Methods, Parameters and Configurations	113
	5.5.4	Overall Evaluation Concept	115
5.6		Summary	118

5.1 Overview

On a very abstract level, the problem addressed within this work can be expressed as illustrated in Figure 5.1: a continuous phenomenon with its specific dynamism in space and time is observed by a set of measurements of particular density and distribution. From these discrete observations of the phenomenon, a continuous model can be derived by applying an interpolation method. This model needs to be discretised for interpretation or analysis. A regular grid of appropriate (spatio-temporal) resolution is much easier to interpret and analyse than the original dispersed observations.

In the context of a *simulation framework* with a synthetic continuous field as phenomenon model—usually realized as a grid—there is the advantage to be able to compare this reference with the model derived from monitoring. The two main processes of monitoring, namely sampling and interpolation, can thus be evaluated by meaningful quality indicators like the root mean square error (RMSE) as the difference between the synthetic model and the interpolated model. By varying methods and parameters of the processes and observing the effects on this quality indicator, the monitoring can iteratively be improved [11, p. 18 f.], [51, p. 114 f.].

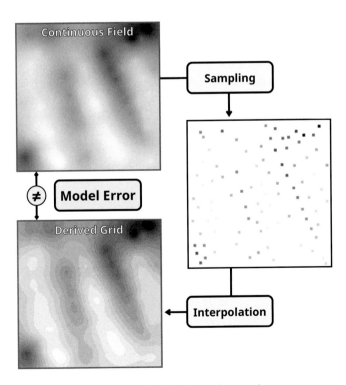

Figure 5.1: Monitoring principle for continuous phenomena.

Besides the quality, monitoring efficiency can be evaluated by introducing indicators for the effort for computation and eventually also for data transmission. By improving methods and algorithms, the expenses in time and energy can eventually be reduced while achieving a similar quality.

As illustrated in Figure 5.1, the main goal of the framework is to allow for continuous improvement of the entire monitoring process according to accuracy and efficiency. The framework presented in this work addresses this goal by (1) creation of continuous random fields and simulation of monitoring, (2) systematic variation of the interpolation method and their parameters (3) evaluation of the process variants using different performance indicators. These concepts and tools will be presented in the remainder of this chapter. Their experimental application and evaluation is carried out in Chapter 7.

5.2 Workflow Abstraction Concept

The area of concern of this work is the acquisition and interpolation of environmental, spatio-temporally referenced observational data (monitoring), the processing of such data (analysis) and the modelling and execution of different variants of these two activities (simulation).

These tasks necessarily include the management and processing of spatio-temporally referenced data, which is often computationally intensive. It is therefore crucial to find working solutions under limited resources, especially for battery operated systems like wireless sensor networks. Besides the efficiency aspect concerning computation time, energy and data volume, the *quality* achieved by the applied interpolation method is the most crucial evaluation metric. It can be used to evaluate several methods and adjust the corresponding parameters to generate best solutions.

In simulated scenarios where the synthetic model provides full knowledge about all relevant environmental parameters, it is easy to determine the quality of the monitoring of a continuous phenomenon by comparing the reference model with the one derived from interpolated observations. In this case, the root-mean-square error (RMSE) is the target indicator to be optimised.

In distributed environments, the transmission of data is often a critical aspect because it is relatively energy intensive. So compression and decompression of data in this context is an important issue (see Section 5.4.3). This is also the case for long-term archiving in databases, where an appropriate indexing strategy is indispensable especially for spatial, temporal and spatio-temporal data for efficient retrieval [17, 107, 111].

On an abstract level, the considerations above can be condensed to a data process model as sketched in Figure 5.2. The model entails the processing unit itself, the input and output datasets, and properties associated with all of those elements.

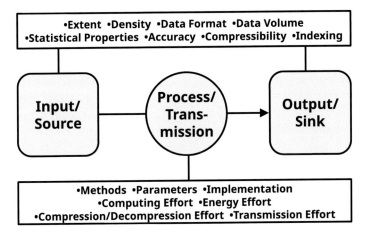

Figure 5.2: Abstraction of a process/transmission step with associated properties.

Taylor et al. [119] use the term *component* for such a processing unit and define it as follows:

> A software component is an architectural entity that (1) encapsulates a subset of the system's functionality and/or data, (2) restricts access to that subset with an explicitly defined interface, and (3) has explicitly defined dependencies on its required execution context.

In the context of the system introduced here, a component's dependencies to the entire system are given by the input data, the output data, the parameters that control its behaviour and the resources necessary to execute it. A complex simulation will be composed of multiple such process steps or components sequenced by their logical order (see Figure 5.3).

Basically, Figure 5.2 entails the generic properties of input and output datasets (source and sink for transmission processes) that affect such a process step. The process itself is determined by its concrete realization (method, implementation, parameters) and the input dataset. To evaluate the quality and efficiency of the process or the transmission, respectively, the indicators for expense in computation, energy, compression/decompression and transmission are identified.

As will be shown in Section 5.5.2, the computational cost for a particular workload can be expressed in terms of time and energy by assigning a specific hardware configuration.

From the data perspective, we find the properties space, time and value, the amount and distribution (point data), the resolution (raster data), and their compressibility and format. Statistical properties can help to decide whether the data can be used for the intended purpose. In simulated scenarios as in this case, it is also possible to exactly quantify the accuracy of the entire monitoring process by indicators like the RMSE. In order to enhance the performance of data retrieval, an indexing can be attached to the data.

With regard to complex computing systems for monitoring, analysis or simulation that need to work with limited resources (computation capacity, time, energy), such abstraction is necessary to evaluate scenarios with respect to different hardware configurations, algorithmic methodologies, corresponding parameters and balancing of workloads in distributed environments.

Having (near) real-time and/or mobile monitoring applications in mind, the factors computation workload, data volume and compressibility (see Section 5.4.3) gain more importance. Given the aspects associated with each process step as depicted in Figure 5.2, a careful balancing of these partly interdependent factors is essential to address both the requirements and the restrictions of a monitoring environment.

The properties will be discussed more thoroughly in the following two subsections.

5.2.1 Datasets (Input/Source and Output/Sink)

There are several generic properties of datasets that appear relevant in the context of a monitoring environment, as shown in Figure 5.2.

The content of a dataset defines its spatial and temporal expansion in a global reference system. It is the crucial criterion to organize extensive environmental data. Without specific indexing techniques, it would not be possible to provide efficient retrieval of the data [107, 6].

The frequency and distribution of observations define the data density for vector data, and the resolution expresses this property for raster data.

The syntactic structure of each dataset is determined by its data format. The underlying model reflects the level of abstraction [47, p. 69] of the described phenomenon.

The data volume that is necessary for each dataset depends on the data format and the number of features (vector) or on the extent, resolution and color depth (raster), respectively.

Compressibility is the ratio by which the data volume can be reduced by applying a compression algorithm. It can produce lossless or lossy representations for both raster [49] and vector [58] data formats.

Statistical properties are of high value when reviewing and analyzing datasets [42]. Classical aggregates like mean and variance provide basic characteristics of the data. More sophisticated indicators like a geostatistical variogram convey deeper knowledge than the general structure of the data. This knowledge can be exploited by applications or users, for example to decide whether a particular dataset has to be considered at all in a particular scenario.

The accuracy of a dataset that represents a field and was generated from observations can be derived by different methods. A root-mean-squared error (RMSE) can usually only be calculated when some reference is given, as is the case for simulations. Cross-validation is often the method of choice for empirical data where the only available knowledge consists of the observations themselves, although it does not necessarily have to be a good accuracy indicator in every case [95, p. 68].

Spatio-temporal indexing of a dataset is the prerequisite for efficient data retrieval [17]. For observational data (vector), the conventional method of defining indexes per feature table might be used. However, the management of these data on the granularity level of single observations might add too much overhead, especially when considering long term archiving.

Instead, it appears reasonable to conflate spatio-temporal areas of observations and exploit their proximity of coordinates and observed values for compression (see Section 5.4.3). The spatio-temporal indexing would refer to those conflated sets which then have to be decompressed on demand. When configured appropriately, this overhead should be outweighed by the benefit of less storage space requirement.

5.2.2 Process/Transmission

On an abstract level, a process step generates an output dataset from an input dataset by applying an algorithm with associated methods and parameters (see Figure 5.2).

Limited resources like computational power, time and energy put considerable demand on the processes to be as efficient as possible. There are generally two different modes of improvement: (1) optimising procedures that are sharply defined according to their result (e.g. sorting of a list) and (2) optimising procedures which are only vaguely defined (e.g. interpolation of observations, fitting of a variogram). In the second mode, there is always a trade-off between cost and effect of a particular procedure that might be difficult to weigh. It is this mode that the present work focuses on.

A thorough analysis of requirements, realistic workloads, appropriate hardware and feasible variants of transmission and processing is necessary to evolve the monitoring towards more and more efficient solutions. Especially environments with wireless communication, big datasets and/or real-time requirements put considerable constraints on the way a task is processed.

A continuous overall optimization requires both the evaluation of the quality of the resulting interpolated model, often expressed by the RMSE, and the tracking of workloads according to transmission and computation (see Section 5.5.2). These indicators have to be registered for each process step and summarized in order to weigh the quality of monitoring against its costs (see Figure 2.2).

For an extensive experimental study which compares various configurations, it is helpful to carry out these variations in a systematic and automated way. Especially when there are manifold methodological and parametric settings that need to be tested and evaluated (see Table 7.4) (see Table 7.7, p. 133?), such an approach can become indispensable for reasonable investigation.

In the context of a complex monitoring scenario as introduced here, there are generally two modes of variation, which can be related to different scales of measure [26], [85, p. 16]. These are respectively appended in brackets:

1. Switching between algorithms and different implementations (nominal scale)
2. Adjusting a parameter (ordinal, interval, rational and cyclic scale)

For the first mode, switching between different variogram models (exponential, spherical, Gaussian) is an example of its application to a simulation scenario (see Section 7.2). The second mode can be used to vary the number of observations by defining a minimum, a maximum and an incremental value (see Section 7.1). This mechanism can also be applied to floating point parameters which are not included in the Gauss-Newton optimization (see Section 5.3.5).

The circular design or "closed loop" [116]—as described in Section 1.2 and depicted in Figure 1.1 (p. 7)—facilitates continuous optimization of monitoring by processing multiple simulation scenarios with different conditions according to sampling design, algorithms and parameters.

Such an optimization can be carried out with respect to several target indicators, of which the following are of central interest with respect to algorithmic improvements:

- quality (e.g. quantified by RMSE)
- logical computational workload (instructions)
- physical computational workload (time and energy)

Whereas the quality indicator RMSE is rather straightforward in the scenario that is regarded here, the computational workload can be either regarded from a *logical* or a *physical* perspective. This differentiation is necessary when the execution effort is to be estimated for different hardware environments. The general concept is introduced in Section 5.5 and is applied experimentally in Section 7.5.

5.3 Monitoring Process Chain

In the last section, the properties of process steps, input and output datasets and performance indicators have been set out. In combination with a solution for parameter variation, a generic toolset for systematic improvement of monitoring is provided.

In this section, each step within a monitoring scenario is specified according to its methodology, parameters, input and output data. An overview of this process chain is given by Figure 5.3.

The main objective in the simulation scenario illustrated by Figure 5.3 is to identify those methods and parameter settings that yield the smallest RMSE (7) and therefore the best approximation of the continuous random field generated by the filter (1). For the sampling of this random field (2), for the generation

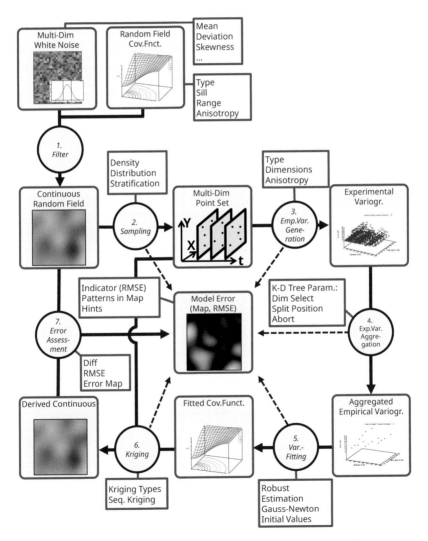

Figure 5.3: Simulation framework architecture with datasets/models (rounded boxes), processes (circles) and their parameters (blue boxes), and their impact on the model error (dashed arrows).

of the experimental variogram (3), for the aggregation of the latter (4), for the variogram fitting (5), and finally for the interpolation by kriging (6), there are multiple variants of algorithms and associated parameters to be evaluated.

The proposed simulation framework was implemented in the programming language C# [93]. With the GNU project *gstat* [52], there already exists a powerful package for geostatistical processing. It is implemented in the statistics-centric programming language *R*.

For the simulation framework that is referred to in this work, the full-featured programming language C# was preferred due to its expressiveness through the support of multiple paradigms, its mature state and wide support, and its portability to almost all platforms. Although this decision means "reinventing the wheel" in many respects, it provides maximum independence and flexibility according to modelling, optimization, portability and interoperability.

In the following subsections, each step of the process chain of Figure 5.3 is specified in detail.

5.3.1 Random Field Generation by Variogram Filter

In order to evaluate different variants of monitoring continuous phenomena, a continuous field is generated as reference model on which sampling and interpolation is carried out. Because continuity is rather a theoretical concept, the field has to be discretised in some form. A regular grid raster as the most common representation of such data structures is also used here.

Beside the two-dimensional grid raster that can easily be visualized as a grey scale image, three-dimensional fields are also used to represent models that include the temporal dimension. Such a model can then be visualized as a sequence of images or a *movie* [131].

These fields are considered, at least approximately, stationary, which means that their statistical properties mean, variance and autocorrelation are invariant under translation in space and time [31]. In the strict sense, however, stationarity is a concept that can only occur in fields of infinite extension (see Section 1.5, also [129]). But since natural phenomena cannot fulfil this criterion either, the data generated by the filter is considered to be sufficiently stationary for the purpose of simulated monitoring.

Pure White Noise Grid

As a prerequisite for a multidimensional continuous random grid that has to be generated, a grid of pure white noise of the required dimensionality and resolution is created. Its grid cells are independent and identically distributed (IID) random values. For the fields generated here, they are characterized by a normal distribution, the preset mean value μ and the standard deviation σ. In order to create normally distributed variables from uniformly distributed pseudo-random numbers, the well-known Box-Muller algorithm [108, 104] is used.

Figure 5.4: Pure white noise grid.

An example of such a random grid is given by Figure 5.4 where it has been applied for two dimensions and transformed to greyscale levels.

When neglecting the concept of stationarity, deliberate variability in mean, variance, skewness, kurtosis or even higher moments can be incorporated in the random grid. This can be achieved by making the parameters of the probability distribution a function of position in space, time or space-time. The approach could be implemented by using a continuous function of position or a pre-calculated continuous surface to control one or more of the parameters.

The resulting continuous but inhomogeneous field could then be used to test the capability of the applied interpolation method to cope with such geostatistical anomalies. The present work, however, is limited to the simple case of the parameters mean and variance which remain constant and thus produce a homogeneous model.

Covariance Function Filter

The general concept of the theoretical variogram and the associated covariance function has been described in Section 3.3. As already mentioned, the covariance function is used to estimate the parameters of the observed field (see Section 5.3.5), to perform the optimal interpolation (see Section 5.3.6) and for the generation of continuous random grids, which will be described here.

The principle of ceasing correlation, as is immanent to any covariance function, is applied for the moving average filter in order to generate a continuous random field from the pure white noise field. Its application to a two-dimensional field is depicted in Figure 5.5(b). The moving average filter—also called mask, kernel or template for two-dimensional grids [49]—defines a value for each cell by which the underlying cell (of the grid it is applied to) is to be weighted. The weight is determined by the associated covariance function and the (euclidean) distance of that cell to the centre of the filter.

For practicability, the filter grid has the same dimensionality and resolution as the white noise field grid it is applied to. In the case of a spatio-temporal grid, each particular cell can be specified by its *spatial* (euclidean norm) and its *temporal* distance to the centre. The result value given by the associated spatio-temporal covariance function is the weight for that filter cell. Due to the identical dimensionality and resolution, the filter can be applied to a target grid by simple matrix-based translations.

Figure 5.5 shows a continuous random field generated by applying such a filter to a white noise field.

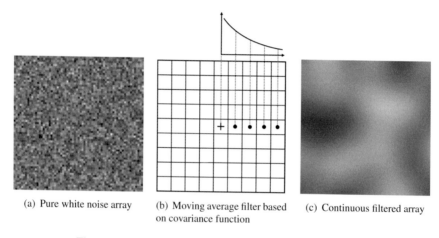

(a) Pure white noise array   (b) Moving average filter based   (c) Continuous filtered array
                       on covariance function

Figure 5.5: Random field generation by moving average filter.

Depending on the applied covariance function, the filter grid has different extensions. If, for example, a *spherical* covariance function (see Figure 3.4, p. 43) is used for filter definition, the correlation between observations further apart than *range* is always zero. Therefore, the grid size of the filter does not need to extend more than corresponding distance for that dimension. Whereas for an *exponential* covariance function where the correlation becomes tiny but never zero, even for large distances, the filter consequently needs to cover all cells of the grid of white noise it is applied to. This means a filter resolution of $2r - 1$, where r is the resolution—for that dimension—of the random field to be generated.

To avoid critical workloads for random grid generation caused by this constellation, an optional restriction is included. The *reach* (not the *range*!) of the covariance function can be restricted to a distance where e.g. less than 1% of full correlation is left. Since the random grid cells affected by these peripheral filter cells are relatively high in amount and relatively low in derived weight, they tend to sum up to zero (relative to the mean value) and can therefore be neglected.

Depending on the size of the filter grid and the current filter position, there is a considerable amount of filter cells that lie outside the target grid. Consequently, they do not contribute to the average value assigned to the target cell on which the filter centre is currently positioned. The proportion of outside filter cells increases towards the fringes and even more towards the corners of the target grid, also depending on the dimensionality.

In some cases, this situation can be avoided by extending the white noise field by $\frac{n-1}{2}$ when n is the resolution of the filter grid in the respective dimension [94]. This approach was not considered here since no "fringe-effect" of a strikingly different pattern could be identified in the generated fields. Furthermore, it would put considerable burden on the random field generation process, especially for filter grids of large relative extension.

Result: Continuous Random Grid

As a product of the statistical operation of a moving average covariance-weighted filter on a pure white noise grid, the continuous random grid has properties that are determined by this process. Since the process of filtering basically generates weighted mean values of the surrounding random cells, the derived filtered grids—at least in tendency—share their mean value with the white noise field used to generate them. From the configuration of the variogram based filter (Figure 5.5(b)), the variance of each cell value of the filtered grid can also be derived as the variance of the weighted sample mean with

$$\sigma_{\bar{x}}^2 = \sigma_0^2 \sum_{i=1}^n w_i^2, \tag{5.1}$$

where σ_0^2 is the variance of white noise field that is equal for each cell and the w_i are the weight values derived from the covariance function for each position in the filter grid.

The value of $\sigma_{\bar{x}}^2$—as being derived from the white noise field and the filter grid configuration—determines the variance of each single cell of the random grid. This value is assumed as an approximate dispersion variance and is therefore used as the initial value for the variogram fitting procedure (see Section 5.3.5), although it is not to be confused with the *sill variance* in the strict sense [129, p. 102].

5.3.2 Sampling and Sampling Density

The sampling design has to be sufficient with respect to density and distribution to capture the underlying phenomenon in a way that adequately addresses the problem or question at hand [22]. Some general issues about the effectiveness and efficiency of sampling have already been mentioned in Section 2.3.2. These considerations will be concretised in the following.

Within a monitoring scenario, sampling is the most fundamental and often also the most expensive task; all subsequent process steps must rely on this limited data about the real phenomenon that is being provided by sampling. The following aspects have to be considered carefully in this context:

- the phenomenon itself and its properties (dynamism in space and time, periodicities, isotropy and trends)
- the sampling design (density and distribution, effectiveness and efficiency of observations)
- the sensor accuracy
- the appropriateness of the selected interpolation method
- the problem to be solved or the question to be answered

When regarding a monitoring scenario as a whole, it turns out that these aspects are not independent but relate to each other, as the following examples show:

- An increased dynamism of the phenomenon makes a higher sampling rate necessary
- Choosing an appropriate and elaborate interpolation method can help to reduce the number of necessary observations
- The more complex a process is, the more observations are usually necessary to generate a model that adequately represents its dynamism
- A dense network of observations is necessary to gain sophisticated knowledge about a phenomenon and thus helps to refine the associated models
- The better the physical processes are understood, the lesser observations will eventually be needed to generate an appropriate model
- Changed requirements according to the aims of the monitoring—e.g. detailed reconstruction instead of rough aggregation—will probably affect the overall effort that is necessary for the monitoring process

In many cases, the dynamism of the continuous field is only roughly known in advance and is therefore not revealed until the data is processed. In the case of geostatistics, the experimental (see Section 3.2) and the theoretical (Section 3.3) variograms derived from the data will contain hints on whether the chosen model is appropriate. It is then up to the operator to decide if the sampling and/or the interpolation model needs to be improved. The monitoring framework might provide suitable indicators—e.g. residuals from the variogram fitting—to support such decision processes.

Beside the "reconstruction" of a continuous phenomenon in space and time—e.g. as a representation of its current state—monitoring can also include the task to check the model for predefined critical states and fire an alert when such a state is present (see Chapter 6, Figure 5.13, p. 90, also [80]).

Such a critical state could be defined by an exceeded threshold in the simple case. For more elaborate applications, it might be formulated as follows: "We need to make sure that we are 95% confident that the nitrogen oxide pollution of district A is below $40\mu g/m^3$ on an average per day." Therefore, it is not sufficient to rely on the interpolated values alone; their confidence interval estimation also needs to be considered here. Only the combination of value and confidence interval estimation will provide enough information to either confirm or reject the presence of a critical state, which, by this particular definition, could also be caused by insufficient sampling. Since the method of kriging explicitly comprises confidence interval estimation, for that reason alone it is an appropriate solution for problems similar to the one described above.

Sampling Density

As already mentioned above, the sampling density that is necessary for an appropriate monitoring of continuous fields depends on its dynamism in each dimension. In practice, this dynamism is either known from previous or similar monitoring scenarios or has to be derived directly from the data (see Sections 3.2, 3.3, 7.2).

The issue is identified by Sun & Sun [116, p. 22] as a "data sufficiency problem", which should be addressed by an "optimal experimental design (OED)". It strives for a good compromise between information content and cost. Meyers [89, p. 187] refers to this problem as "representativeness". Consequently, in the realm of monitoring continuous phenomena, some estimation of when a sampling density is *sufficient* is necessary.

Within the proposed simulation framework, the dynamism can be determined by setting the spatial, temporal or spatio-temporal parameter(s) sill and *range* of the variogram filter used to generate the reference field (see Section 5.3.1). These parameters can then be compared with the ones derived from the variogram fitting procedure applied to the simulated observations (see Section 7.2).

The main objective that is addressed here is to quantify the relation between this dynamism and the average sampling density that is necessary to capture the phenomenon adequately. Instead of applying heuristics like "nested survey" [129, p. 127] in order to systematically inspect the autocorrelation structure of a particular phenomenon, we will try to find some law, or at least some rule of thumb, to derive the necessary average sample rate from the extension and the dynamism of the phenomenon. If this rule is valid for synthetic fields, it is assumed to be applicable to real-world phenomena for which the dynamism is estimated by the parameter *range* for each dimension.

To approach this problem, it is first reduced to the one-dimensional case in order to look for analogies to the Nyquist-Shannon sampling theorem which is well known in signal processing [101].

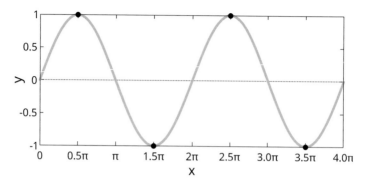

Figure 5.6: Nyquist-Shannon sampling theorem.

As can be seen in Figure 5.6, according to the theorem, at least two samples per wavelength are necessary to capture a periodic sine signal. The sampling distance *d* is thus determined by

$$d = \frac{\lambda}{2}. \tag{5.2}$$

Since periodicity is usually not found in natural continuous fields, the sample rate necessarily needs to be higher to capture such phenomenoa appropriately. For approximation, uniformly and randomly distributed samples instead of systematic or stratified ones [22] are presumed, because such regular configurations are hardly encountered for mobile wireless sensors [121]. Furthermore, systematic or stratified distributions will reduce the variety of small distances which are crucial for variogram estimation [95, p. 53], [8, p. 53].

Two general challenges need to be addressed to transfer the principle of Nyquist-Shannon to the domain of multidimensional continuous phenomena:

1. Finding a reasonable factor to relate the wavelength of a periodic signal to the parameter *range* of general continuous phenomena
2. Extending the principle from one-dimensional to multidimensional applications

In order to obtain an estimate of the geostatistical concept of *range* within a sine signal, the experimental variogram for 100 uniformly distributed observations within one wavelength of a sine function is generated. From this, the semi-variance can be derived by Equation 3.2 for each pair of observations. The experimental variogram is generated by plotting these values against their corresponding pair distances (see Section 3.2). As can be seen in Figure 5.7, the experimental variogram for the 4950 possible pairings from 100 uniformly dispersed observations converges to zero as the distance approaches the value of 1 (unit: fraction of the wavelength of 2π) and is confined by sine shaped upper

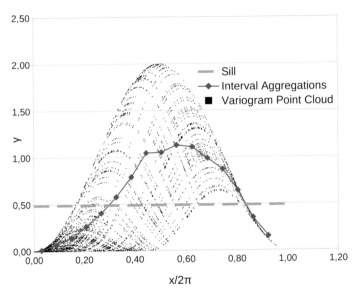

Figure 5.7: Experimental variogram with semivariances (γ) plotted against pair distances (normalized to the wavelength of 2π) of observations on the sine signal; the *sill*, represented by the green horizontal line, is intersected by the polygon of aggregated interval points; the abscissa position of this intersection is considered as *range*.

and lower bounds. These regular patterns are caused by the periodicity and are usually not found in experimental variograms. However, this can be neglected here since we are only interested in the *first* position from which the dispersion of values exceeds the total variance.

As common in geostatistics, the trend of the variogram can be approximated by interval-wise aggregations of the experimental variogram points [129, 52]. The polygon connecting those aggregation points represents this trend. It is this geometry to which a theoretical variogram is usually fitted by adjusting its parameters (and therefore the parameters of the covariance function, see Section 5.3.5).

In the case of the sine signal, however, there is no appropriate theoretical variogram to fit to since the semi-variances γ decrease when approaching the value of 1.0, which is not the case for a valid variogram. Furthermore, since the only value of interest is *range* here, there is quite a straightforward way to roughly estimate it.

As can be seen in Figure 5.7, an approximation of the value *range* can be derived from the first point of intersection between the total variance (or dispersion variance [129]) of the dataset (horizontal line) with the polygon line.

This point is assumed to represent the threshold distance between point pairs from which the dispersion (or semi-variance) between the point values is just as large as the total variance of the dataset. As already mentioned in Section 3.3, this does not strictly comply with geostatistical practice, but is considered to be sufficient to derive an approximate value for the minimum sampling density.

The experiment as depicted in Figure 5.7 was repeated 30 times and on average reveals a dispersion (or total) variance of 0.4878 with a standard deviation of 0.0503. The average ratio between wavelength λ and range r is 0.2940 with a standard deviation of 0.0188. For convenience, this value is rounded of to the safe side (down), so the range-wavelength ratio is estimated by

$$\frac{r}{\lambda} \approx \frac{1}{4}. \tag{5.3}$$

To adequately capture a sine-shaped signal for interpolation by the method of kriging, we assume a minimum average coverage by two observations per range distance (or eight observations per sine wavelength, respectively) because this is the minimum number of observations to at least *detect* a correlation above the average correlation within one range distance. From that, the sampling distance d_p can be derived by,

$$d_p \approx \frac{\lambda}{8} \tag{5.4}$$

to capture *periodic* signals of wavelength λ for kriging interpolation and

$$d_c \approx \frac{r}{2} \tag{5.5}$$

for *continuous* non-periodic signals with range r.

To derive the number of samples for regions of arbitrary extension, we need to apply a factor f that represents how many times the (average) sampling distance d_c is contained within the extent e by

$$f \approx \frac{e}{d_c}. \tag{5.6}$$

Together with Equation 5.5, we can now determine the number of samples c necessary per dimension i by

$$c_i \approx 2\frac{e_i}{r_i}. \tag{5.7}$$

The fundamental relationship between a continuous phenomenon of range r, the extent e and the number of observations c_1 necessary to capture it, is thus defined. It can be used to estimate sampling density for *one* dimension. To generalize the concept in order to be applicable to multiple dimensions, we calculate the product of its n dimension-wise representatives by

$$c_n \approx \prod_{i=1}^{n} 2\frac{e_i}{r_i}. \tag{5.8}$$

This expression is used to estimate the minimum number of uniformly distributed random samples on multidimensional continuous fields as a function of their values of extent and range for each dimension. Thus, we can determine an appropriate sampling density for arbitrary initial configurations of *sill* and *range* in the random reference model. The approach is experimentally validated in Section 7.1.

Result: Multidimensional Point Set

Depending on the simulated or actual process of sampling, the set of observations represents the degree of knowledge about the observed phenomenon that is available. The geostatistical properties (mainly the parameters *sill* and *range*) of the theoretical variogram are not necessarily equal to the ones within the observed region, therefore it is also called the *regional variogram* [129]. In practice however, the properties derived from the observations are often the only ones available and thus have to be worked with.

For a good estimation of the geostatistical properties of a field, the distribution of observations should sufficiently cover all distances relevant to the given problem to provide enough information for variogram fitting [22, 129]. Especially the short distances are of decisive importance for variogram estimation [8, p. 53], [95, p. 53].

With real world data, there might be anomalies like anisotropy that will usually materialize in the derived experimental variogram cloud. If no other geostatistical property information is available (e.g. from previous samples), the set of observation points is a carrier of both (i) the discrete spots of knowledge about the phenomenon to be interpolated between and (ii) the statistical properties this interpolation has to be based on [125].

5.3.3 *Experimental Variogram Generation*

The experimental variogram has already been introduced as a basic geostatistical concept in Section 3.2 and was also applied to determine the *range* property of a sine-shaped signal in Section 5.3.2.

Given a sufficient number of observations by following the method as proposed in the previous section, we will now set out the process steps that are necessary to derive the parameters of the theoretical variogram from them. Namely, these steps are (i) the generation of the experimental variogram (this section), (ii) the aggregation of the variogram points (Section 5.3.4) and (iii) the variogram fitting (Section 5.3.5).

For better visual demonstration of the method, we stick to a two-dimensional continuous random field generated with following parameters:

- white noise field: 150 x 150 grid cells, *mean* = 5000, *deviation* = 500
- variogram filter: separable gaussian, range of 75 grid cells, resulting grid cell value deviation of 4.65
- sampling: 20 points (derived by Equation 5.8), uniformly distributed
- experimental variogram: 190 variogram points comprising of spatial distance and semi-variance γ derived from the sample point pairings
- aggregation: 16 aggregates (by Equation 5.10 with $b = 1.5$, $c = 0.8$), partitioning dimensions aggr. method: median

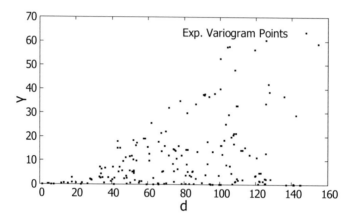

Figure 5.8: Experimental variogram point cloud.

As can be seen in Figure 5.8, the semivariances (ordinate) tend to scatter more with increasing pair distance (abscissa). A spatio-*temporal* variogram [31, 52] can be visualized as a three-dimensional plot with spatial distance, temporal distance and semi-variance as axes (see Figure 3.1).

Although the number of variogram points would also allow for direct fitting of the theoretical variogram in this example, we will apply binary space partitioning (BSP) as an aggregation approach, since its principle is more comprehensible with small datasets. It will be introduced in the next section.

5.3.4 Experimental Variogram Aggregation

Before the interpolation by the geostatistical method of kriging can actually be carried out, a formal description of the spatio-temporal autocorrelation is needed. After generating the experimental variogram as a point cloud in the previous step, the theoretical variogram function associated with this covariance function (see Section 3.3) has to be fitted to this point cloud [92, 18, 53]. The number of variogram points in the experimental variogram n_v depends on the number of samples n_s by

$$n_v = \frac{n_s^2 - n_s}{2}. \tag{5.9}$$

The subsequent and rather complex step of variogram fitting can therefore become too expensive for large datasets. The common solution for this problem is to perform some aggregation that retains the general dispersion characteristics of the original variogram points [129].

Depending on the dimensionality of the observational data and the dimensionality of the associated variogram model (spatial/temporal/spatio-temporal, isotrop/anisotrop), the aggregation of points in the respective experimental variogram will have to be possible with various dimensionalities to be generic.

The common structure of the experimental variogram for all dimensionalities is that of n independent variables (e.g. spatial and temporal distances between pairs of observations) and one dependent variable, which is the semivariance γ as given by Equation 3.2.

This is also the target variable of the theoretical variogram to be aggregated by using the spatio-temporal proximity of points as criterion for grouping. Therefore, the common concept of aggregating the original variogram points by using intervals of constant lag intervals [129] is extended (or generalized) by the following features:

- Instead of using only one dimension for segmentation of the experimental variogram, binary space partitioning (BSP, [111]) allows for multidimensional segmentation
- Instead of rigid segmentation (e.g. by constant interval size), more flexibility is achieved by variable hyperplanes that can adapt to the given data structure

This approach aims at a generic and robust solution for the central problem of geostatistics: variogram fitting. It provides flexibility in terms of

- the dimensionality of the used variogram model
- the number of points of the experimental variogram
- the dispersion of points

As already mentioned, the experimental variogram is a set of points in $\mathbb{R}^d$, where $d - 1$ dimensions represent parameters of the particular variogram model and one dimension represents the target variable γ. For aggregation, all dimensions except that of γ are used for binary space partitioning, thus generating disjunct regions with subsets of the original variogram point set.

According to its common purpose of space partitioning for search operations, a BSP tree usually—e.g. when implemented as k-d tree—subdivides a set of objects into two subsets of equal number of elements. In a recursive manner, the partitioning dimensions or axes do cycle according to a predefined order [111].

For the sake of adaptivity, the partitioning method is extended to let diverse statistical parameters control the process. Figure 5.9 illustrates critical decisions within the BSP tree algorithm where statistical properties are used to determine how the partitioning is carried out. Therefore, the algorithm keeps track of the statistical properties of each dimension separately. So for each set of points the minimum, maximum, extent, mean, median and variance are calculated per dimension and can thus be used as control parameters for the points of decision that are described below.

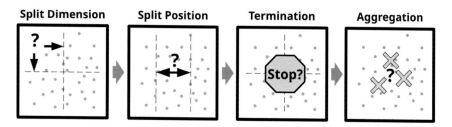

Figure 5.9: BSP tree partitioning process: options for control by statistical properties.

Split Dimension

If there is more than one free variable in the variogram, as is the case for spatio-temporal models, the recursive partitioning algorithm needs to select the next splitting dimension, or, in other words, which coordinate axis will be the normal vector of the next splitting hyperplane.

The statistical properties described above can be used to determine the next dimension. So it might be reasonable to select the dimension with the greater extent or deviation for the next split. In many cases it is appropriate to relate this value to the one of the total set that the procedure started with to get a relative value. This variant was also applied here.

For a uniform splitting pattern, the algorithm can also simply toggle between all dimensions without any parameter checking. This is the behaviour of a standard k-d tree [111].

Within this solution, consecutive splits by the same dimension are allowed, which might be useful for very anisotropic point distributions. More complex definitions where several parameters and conditions are combined might also be reasonable to adapt to such data structures, but they are not regarded here.

Split Position

Once the dimension for the next split is determined, the position of the hyperplane on the corresponding axis has to be set. The statistical properties of this particular dimension can be used for the determination of this position.

In the case of the median value as split position, subsets with equal numbers of elements are obtained, which is known from k-d trees [111]. Alternatively, the mean value can be used to give outliers more influence than with the robust median. Simply selecting the middle position will result in a regular grid, but only when split dimensions toggle and tree depth is equal for each dimension. All of these variants are tested and evaluated in Section 7.2 (see Table 7.1, p. 135).

Termination

There are several conditions that can be used within a BSP tree algorithm to terminate the recursive partitioning. In the context of aggregation of the experimental variogram, the following are considered reasonable:

∎ maximum tree depth
∎ maximum elements per leaf
∎ maximum total number of leaves
∎ spatial extent of current leaf

Each of these variants has its advantages and drawbacks: A constant maximum tree depth is easy to implement, but does eventually not adapt well to the given data structure. A constant maximum spatial extent of leaves is also straightforward, but may produce subsets with numbers of elements differing by too much. A constant maximum number of leaves total is difficult to implement in a recursive manner if the tree does not become too unbalanced.

In order to achieve robust behaviour, more complex termination rules can be defined by the logical combination of multiple conditions.

For this study, the termination condition of maximum elements per leaf was implemented. From the algorithmic perspective, this condition is fulfilled when stopping the recursive partitioning as soon as the threshold number of elements is achieved or undercut. The condition produces statistically similar subsets of points to be aggregated. For reasonable sizes of those subsets while given an arbitrary total number of elements, the logarithm-based formula

$$n_a = c \cdot log_b(n_t) \tag{5.10}$$

is used, where n_a is the total number of aggregated points to be created, n_t is the number of points in the original variogram, b is a logarithmic base that controls the degree of decreasing, and c is a linear scaling factor. By applying this formula, an arbitrary choice of the number of aggregations is avoided. It adapts to the total amount of original variogram points by producing reasonable and feasible numbers of aggregated points.

Aggregation

The preceding procedure provides n datasets of the original experimental variogram dataset, separated by BSP hyperplanes. To actually *aggregate* these sets to one point for each of them, there are several options taken into account:

■ mean value
■ median value
■ middle of the corresponding BSP tree partition interval

These options can be assigned individually to each of the independent dimensions used for BSP tree partitioning. For the target variable γ itself, only the mean value is assumed to aggregate the dispersion correctly [95, p. 16], [30, p. 59]. Also these variants are tested in Section 7.2 (see Table 7.1, p. 135).

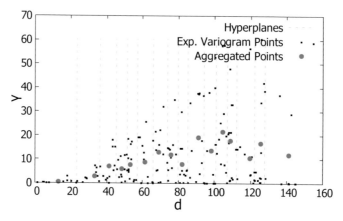

Figure 5.10: Aggregation of variogram points; different interval sizes result from adaptive BSP algorithm.

Figure 5.10 illustrates the BSP aggregation that is applied to the point cloud from Figure 5.8. Since the points of the experimental variogram represent the statistical properties of the dataset, the aggregation is supposed to be carried out in a way that transmits, at least approximately, the significant properties of the original point cloud to the reduced point set.

As can also be seen from the plot, the aggregated set of points is by far less dispersed than the original variogram point cloud and already indicates a continuous function. Beside the geometrical properties, each aggregation produces additional statistical data like variance or skewness that could be used to define weights for the subsequent variogram fitting process step of [52]. However since the aggregations are already statistically similar due to the termination condition of maximum elements per leaf, this mechanism is not considered here.

5.3.5   *Variogram Fitting*

The aggregation of the original experimental variogram generates a dataset of reasonable size for the adjustment of the parameters of a theoretical variogram. Because of the redundancy of data points, they will not fit the theoretical variogram without residuals and a non-linear optimization method like Gauss-Newton has to be used to estimate its parameters [116, 4].

As already mentioned in Section 3.3, the theoretical variogram is a mirror image of the covariance function [129, p. 55]. So by fitting the theoretical variogram to the points that were aggregated from the experimental variogram, we obtain the parameters of the respective covariance function needed for kriging.

Generally, the problem can be defined by fitting the variogram model $\gamma = f_p(x)$ with x being the distance (spatial, temporal or spatio-temporal) for which the variogram γ is obtained.

The Gauss-Newton algorithm [116, 122] iteratively determines the vector of parameters $p_1, ..., p_k$ that minimize the squared residuals between the observations (here: the aggregated variogram points) and the function values at the respective positions [128, 4]. By equipping each data point with a weight w_i, the process can consider specific circumstances which are supposed to have an influence on the optimization result.

The distance from the origin of the variogram, the number of points used for the preceding aggregation or the variance of their mean value are reasonable parameters to define the weights [28]. Since an approximate number of equal points per aggregate are provided by the BSP algorithm, the weighting is derived from the (n-dimensional) distances from the origin.

Weighting

In order to get a good estimate of the variogram at its origin, the points near the ordinates should be given more weightage than the more distant ones [8, p. 53], [95, p. 53]. A weighted variant of the Gauss-Newton algorithm was implemented to achieve a higher influence of the elements of low n-dimensional distance by defining the weights per aggregated point by

$$w_j = \prod_{i=1}^{n} 1 - \frac{d_{ji}}{max(d_i)}, \tag{5.11}$$

where d_{ji} is the distance of the aggregated point j to the origin regarding dimension i and $max(d_i)$ is the maximum distance that occurs in the whole set of aggregated points. The value for each dimension and therefore also the product, is guaranteed to be between 0 and 1.

A smoother decrease of weight by distance is achieved by the sine-based function:

$$w_j = \prod_{i=1}^{n} 1 - sin^2(\frac{\pi d_{ji}}{2}). \tag{5.12}$$

Alternatively, to achieve a stronger differentiation of weighting between points near to and points far from the coordinate origin, a weighting function based on a variant of the logistic function [106] was applied with

$$w_j = \prod_{i=1}^{n} 1 - 1/(1 + exp(g(1 - 2\frac{d_{ji}}{max(d_i)}))), \tag{5.13}$$

where factor g controls the gradient. It is set to 5.0 in Figure 5.11 which contains plots of the functions for one dimension. The weighting variants introduced here will be evaluated experimentally in Chapter 7.

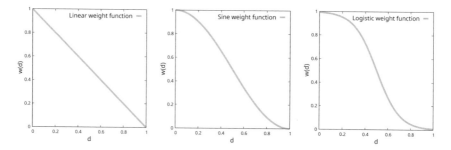

Figure 5.11: Weighting functions relating relative distance to origin (d) to weight (w): (1) linear, (2) sine-based and (3) logistic.

The number of points, the variance, or other parameters from the preceding aggregation process could be used alternatively or be incorporated into the proposed weighting functions to give them some influence on the Gauss-Newton procedure.

This variant was not implemented in the framework because it is assumed that the combination of adaptive aggregation and distance weighting is sufficient to cover this aspect for the present study. But this or similar variants could easily be evaluated using systematic variation and evaluation (see Section 5.5).

Figure 5.12 shows a Gaussian variogram model fitted by linear weighted aggregation points. The parameters which were illustrated in Figure 5.9 are set as follows: split dimension: not applicable here, since variogram contains only one spatial dimension; split position: median; termination: maximum number of points; aggregation: mean. The logistic function was used for the weighting of the subsequent fitting by the Gauss-Newton algorithm.

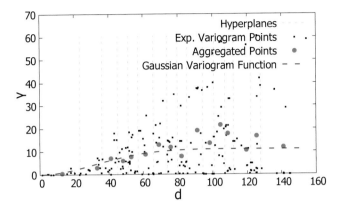

Figure 5.12: Theoretical variogram fitted to aggregated variogram cloud.

As the graph reveals, the algorithm yields a reasonable fitting to the variogram points at sight. The *visual* assessment is still an important issue of variogram fitting [95, p. 38 f.], [8, p. 54]. There are also approaches to completely automate the fitting [100, 34]. However, because of the sheer amount of varieties of kriging and its associated parameters, the task of selecting and fitting a variogram is very complex for which no established solution is available yet (see Chapter 8).

Multiple Initial Values: Hybrid Approach

As is a common problem for non-linear optimization algorithms, the Gauss-Newton method is also not guaranteed to converge for every constellation of initial values [116, p. 63]. The aggregation of the variogram points and the weighting scheme already reduces this risk, but does not completely eliminate it.

To address this problem and to get better results in case of multiple local minima, the optimization procedure is started with varying initial parameter values. The variants are generated by a n-dimensional subdivision of the parameter value or values. Thus, a set of starting parameter variants of size $x \cdot n$ is generated, where x is the number of subdivisions per dimension and n is the number of dimensions of free parameters.

It is not the whole value domain that is used as an initial interval to be subdivided per dimension. Instead, the thresholds are determined by robust estimation based on quantiles.

Given this set of initial parameter settings with predefined criteria when the iteration should cease—for both cases: sufficient converging as well as diverging behaviour—leads to a set of result values with usually different values of residuals.

In the ideal case, all starting parameter variants converge to the same result, which is only the case for very robust constellations. Except for ill-conditioned sampling constellations, this approach provides a robust estimate of the parameters with minimum residuals. Local minima are more likely to be found this way.

Depending on the sensitivity of the given constellation, the number of initial values can be of decisive importance for achieving an optimal solution. With too many variants, however, this complex process might exceed the computational capacities.

In the experiments carried out in Section 7.2, a moderate amount of variants was sufficient to yield feasible solutions. For situations where this is not the case, more sophisticated methods to improve convergence should be applied [5, 116, 113].

5.3.6 Kriging

Once the appropriate variogram model is determined, the interpolation procedure itself includes the inversion of the covariance matrix (once per model) and its application to determine optimal weights by which each observation contributes to the estimated value (once per interpolation). Based on these weights, kriging also provides an estimated confidence interval for each interpolated point. The general proceeding of kriging has already been set out in Chapter 3.

With big numbers of observations, the inversion of the covariance matrix might produce critical workloads due to its complexity of $\mathcal{O}(n^3)$ [46, p. 503], [116, p. 356]. There are various approaches that address this issue [130, 57, 100, 24, 10, 97].

In this work, an approach is introduced that addresses the problem of computational burden *and* that of continuous integration of new observations into an existing model (see Section 5.4.2). It exploits the estimation variance (kriging variance) that is provided by kriging. This feature makes the method outstanding amongst other interpolation methods.

5.3.7 Error Assessment

Given the same extent and resolution for both the continuous reference random field and the one derived by interpolation of observations, the deviation between the two models can easily be calculated. The RMSE provides a compact indicator for the overall quality of observations and interpolations. The effect of changed variants or parameters (see Section 5.5) can thus be quantified.

An error map or map of the "second kind" [89, p. 464] of the same resolution can help to reveal more subtle patterns indicating systematic flaws of the monitoring process (see Section 7.7).

Representing a single-number summary of the error map, the RMSE itself is given by

$$RMSE = \sqrt{\frac{\sum_{i=1}^{n}(\hat{y}_i - y_i)^2}{n}}, \tag{5.14}$$

where for each grid cell i, $\hat{y}_i$ is the value of the derived model, y_i is the value of the reference model and n is the total number of grid cells of the model.

Unlike the situation in a real world monitoring scenario where interpolation quality has to be estimated by approaches like cross validation [43, p. 147], the synthetic reference model of arbitrary resolution means total transparency of the errors caused by sampling and interpolation. While the RMSE will in most cases be sufficient to compare the interpolation quality of different monitoring process variants, the more verbose representation of the error as a deviation map can provide valuable hints for further improvement.

Geometric patterns within the deviation map that significantly differ from pure random fields can indicate potential for systematic improvement of the monitoring process. So different patterns in the error map might provide visual hints to particular deficits in the monitoring process that produced the model associated with it.

- A predominantly high error value that is only mitigated regionally at the spots around the observations might indicate an insufficient density of samples
- Distinctive border areas of high slope (discontinuities) that separate regions of rather homogeneous error values are a hint for an insufficiently fitted variogram model
- A rather continuous error map with moderate error values at spots of maximum isolation from observations indicates a near optimal monitoring configuration

While the RMSE provides a straightforward quantification of the model quality that can be used to easily identify the best among many solutions, the error map reveals subtle patterns that might contribute to more thorough investigations.

Both approaches, however, do only work with synthetic models or with phenomena that are available at much finer resolution than necessary in the planned monitoring scenario in order to serve as reference for experimental study. In this work, error assessment is applied to both synthetic phenomenon models (see Section 7.2) and real remote sensing data (see Section 7.7) as a case study.

5.4 Performance Improvements for Data Stream Management

Monitoring of continuous phenomena poses several specific challenges according to the processing and the archiving of observations. Some of them that are considered to be crucial are addressed in this section.

Providing an interpolated grid from a set of discrete observations means considerable computational burden if massive data or real-time requirements (or both!) is present. Also, the seamless and efficient actualization of a calculated model by new incoming observations is indispensable for (near) real-time monitoring systems. Both problems are addressed by the approach that is set out in Section 5.4.2.

Although storage costs are continuously decreasing, the archiving of extensive observational data might nevertheless reach critical dimensions. While grid data as derived from interpolation provides better interoperability, retaining the original observational (vector) data has several advantages (see Section 2.2.2). A compression algorithm specifically designed for such data is introduced in Section 5.4.3.

5.4.1 Problem Context

The specific features introduced in the next two sections can best be explained in the context of a monitoring system architecture as sketched in Figure 5.13.

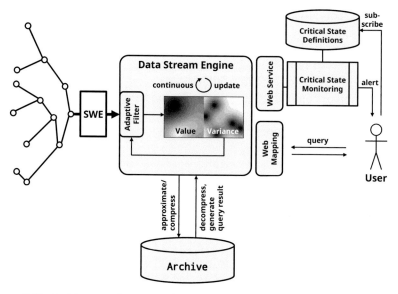

Figure 5.13: Architecture of a system that processes, visualises, monitors critical states and archives sensor data streams.

The envisioned data stream engine (DSE [42]) continuously processes incoming observations (provided by sensor web enablement or SWE) and integrates them into the model that reflects the current state, eventually as a Web Map Service [15]. Beside the value of interest, the model also keeps track of the deviation map as "map of the second kind" [89, p. 464].

As already mentioned in Section 3.5, this deviation map or variance map can be used for several purposes. It can indicate insufficient confidence for critical state monitoring (see Chapter 6). For massive loads of data, it can be used as an adaptive filter to only let non-redundant observations pass. Its role as weighting schema for the merging of sub-models—in order to mitigate computational workload or to support a continuous update by new observations—is the subject of the next section.

From the user's perspective, the monitoring system should provide the model at arbitrary points in space and time within the region of interest. So the data prior to the current model needs to be archived and retrieved appropriately. As indicated by Figure 5.13, the process of compression and decompression should be hidden from the user who usually accesses the data by some (web-) interface.

Based on web services, a data stream engine (DSE) should provide interfaces for both interactive web mapping and automated monitoring. For the latter, critical states can be defined, subscribed to a specific database and checked against the current map regularly. Such definitions can refer to values (e.g. for an alert after an exceeded threshold), confidence estimations (when more measurements are necessary) or both combined (high risk of exceeded threshold [55, p. 24]).

For queries on historical data and for long-term analyses, an archive containing data which is compressed by approximation is maintained alongside the real-time services. When queried, it is decompressed and provided as a digital map.

The methodologies introduced in the following sections are in principle designed to support the functionalities of a monitoring system as sketched above.

5.4.2 *Sequential Model Merging Approach*

Instead of processing a set of observations all at once in a single step, a sequential approach can mitigate some of the associated problems. Being a by-product of interpolation, the kriging variance is exploited for this purpose.

5.4.2.1 *Overview*

As already mentioned, the sequential merging approach addresses two common problems in the context of monitoring continuous phenomena:

1. Reducing the computational workload for big datasets

2. Allowing for subsequent and smooth model updates in data stream environments

The general task that these problems are associated with is to generate a regular grid from often arbitrarily distributed and asynchronously conducted discrete observations. The general interpolation problem is a subject matter of spatio-temporal statistics [31], while the peformance issue is addressed by data stream management [42]. Consequently, the proposed solutions for these problems depend much on which context they are solved in.

5.4.2.2   Related Work

Whittier et al. [131] suggest a new design of a data stream engine (DSE) that is based on k Nearest Neighbors (kNN) and spatio-temporal inverse distance weighting (IDW). It uses main memory indexing techniques to address the problem of real-time monitoring of massive sensor measurements. In contrast to this approach, we want to avoid a sub-model based on a fixed sized temporal interval. By merging sub-models continuously, we also consider old observations if no better information is available. This is especially important when observations are inhomogeneously distributed in space and time.

Appice et al. [6] inspect trend clusters in data streams and discuss techniques to summarize, interpolate and survey environmental sensor data. Since one main application is the detection of outliers within a phenomenon of rather low dynamism, the approach provides a coarse approximation by clusters of similar values. For our purpose, a smooth representation of each state is desirable.

Walkowski [126] uses the kriging variance to estimate a future information deficit. In a simulated chemical disaster scenario, mobile geosensors are placed in a way that optimises the prediction of the pollutant distribution. Instead of optimising the observation procedure itself, we exploit the kriging variance in order to achieve efficient continuous model generation from massive and inhomogeneous data.

Katzfuss & Cressie [65] decompose a spatial process into a large-scale trend and a small-scale variation to cope with about a million observations. This solution is an option for optimizing very large models, but is not helpful for our sequential approach with its real-time specific demands.

Osborne et al. [96] introduce a complex model of a *gaussian process* (synonym for kriging) that incorporates many factors like periodicity, measurement noise, delays and even sensor failures. Similar to our work, sequential updates and exploitation of previous calculations is performed, but rather on a matrix algebra basis. It uses kriging with complex covariance functions to model periodicity, delay, noise and drifts, but does not consider moving sensors.

5.4.2.3   Requirements

Concluding from the state of the problem area as characterized by the related work above, the following requirements are considered to be crucial within the scope of this work:

- Locally confined, smooth and flexible updates of interpolated models
- Preserving confidence estimate (kriging variance) as crucial information also for adaptive filtering and critical state checks (see Section 3.5, Chapter 6)
- Provision of immediate coarse results generated by subsets of observations
- Preserving of preceding computational effort

5.4.2.4 Principle

The sequential merging approach that is set out here exploits the variance map provided by using kriging Equation 3.11. Depending on the purpose, it can also be represented as *deviation* map (see Figure 5.14).

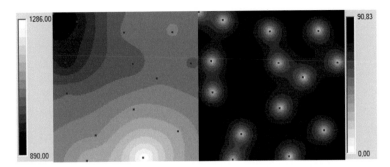

Figure 5.14: Kriging result with value map (l) and corresponding deviation map (r). The red dots represent the observations.

The variance or deviation map represents the degree of confidence in the interpolated value and therefore can be used to calculate how much it should contribute to the resulting value when combined with another model of the same region but with different observations. The principle of this approach is visualized in Figure 5.15.

The approach uses the inverse variances as weights [59] when fusing two grids generated from different sub-sets of observations of a region. When applied sequentially, this method successively "overwrites" the former grid, but only gradually and in regions where the new grid's variance is significantly lower. The variance maps themselves are also fused (eventually taking into account temporal decay), thus representing the confidence distribution of the new model and determining its weighting schema for the subsequent fusion step.

The process is performed for each grid cell by deriving the weight $p_{[i]}$ from its variance with

$$p_i = \frac{1}{(\sigma_i^2)^d},\tag{5.15}$$

where σ_i^2 is the kriging variance of each grid cell, and d is an optional parameter to control the grade of weight decay relative to the variance of the model to be merged with. This factor might be adjusted according to the spatio-temporal dispersion of the given dataset. When set to 1.0, it is simply an inverse-variance weighting [59].

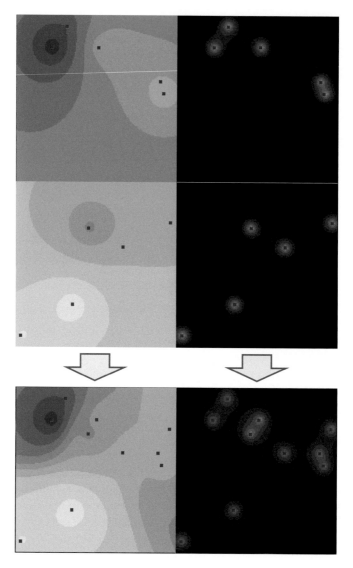

Figure 5.15: Merging of models by using weight maps: the values (l) and variances (r) of two models are merged to a resulting model that combines the information they contain (bottom).

With values and weights for each grid cell, the merged model values x_{i+1} can be derived from the current sub-model values x_i and previous model values x_{i-1} by

$$x_{i+1} = \frac{x_{[i]} \cdot p_{[i]} + x_{[i-1]} \cdot p_{[i-1]}}{p_{[i]} + p_{[i-1]}}. \tag{5.16}$$

Equation 5.16 assumes the merging of two models, which could be applied for a continuous update in a real-time monitoring scenario. For the more general case with an arbitrary number of models, the expression

$$\bar{x} = \frac{\sum_{i=1}^{n}(x_i p_i)}{\sum_{i=1}^{n} p_i} \tag{5.17}$$

provides the weighted result value $\bar{x}$. Its variance can be determined by

$$\sigma_{\bar{x}}^2 = \frac{1}{\sum_{i=1}^{n} p_i}. \tag{5.18}$$

In the case of real-time monitoring where the current model has to be continuously merged with new models generated from new observations, a temporal decay should be applied to the preceding model. A simple exponential decay factor f_d can be applied as

$$f_d = b^{\left(\frac{t-t_0}{r_t}\right)}, \tag{5.19}$$

with $t - t_0$ representing the time elapsed since the last model was generated, and b being the fraction that shall remain after time range r_t. In principle, any other covariance function (see Section 3.3) might be used to define the temporal decay rate.

5.4.2.5 Partitioning Large Models: Performance Considerations

Apart from the continuous update mechanism as assumed above, the proposed method can also be used to partition large models and apply it in a divide-and-conquer manner [23].

Kriging goes along with a high computational complexity—caused by the inversion of the covariance matrix—of $\mathcal{O}(n^3)$ [116, p. 356], [96, 10], where n is the number of samples. Considering this fact in the context of a massive data load in combination with (near) real-time requirements, this can become a severe limitation of the method. Hence, when sticking to its essential advantages like the kriging variance, the merging strategy can be applied to mitigate the computational burden while delivering comparable results.

The original set of observations is separated into s subsets to which the kriging method is applied separately. The resulting sub-model grids are in the same area as the master model that contains all points. To consider all measurements in the final model, the sub-models are sequentially merged with their respective predecessors, as shown in Figure 5.16.

Alternatively, all sub-models might be calculated before they are merged in one step using Equation 5.17. This approach would, however, not provide the advantage of an immediate—albeit coarse—result. Since the linear combination of values is not equivalent to the subsequent variant, the resulting model will

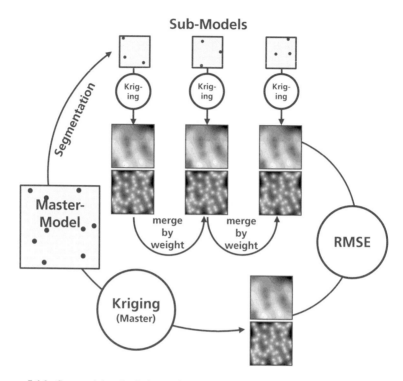

Figure 5.16: Sequential calculation schema: model partitions calculated separately and merged sequentially; the loss of accuracy induced by this approximation is indicated by the RMSE.

also differ. With the applicability for continuous updating of real-time systems in mind, only the sequential approach was investigated further here.

As is the case for any approximate solution, there is a trade-off between performance gain and resulting accuracy. As for other cases in this work, the loss of accuracy is quantified by the Root Mean Square Error (RMSE) against the master model.

In a spatio-temporal context, the segmentation should be performed with respect to the order of timestamps, thus representing temporal intervals per submodel. This also applies to real-time environments where subsequent models are to be created continuously.

For a pure spatial model, the subsets of points can be generated randomly. Here, the order of sub-models does not represent the temporal dynamism of the phenomenon, but rather a utilisation level of information with associated estimated accuracy. This is also the case for the configuration as introduced below.

The segmentation and associated sequential calculation limits the potential complexity of $\mathcal{O}(n^3)$ to the size of each subset s. This can be set as a constant, but could also be dynamically adaptive to the data rate. In any case, there should be an upper bound for the size of sub-models to limit the computing complexity.

While doing so, the merging procedure itself can be costly, but grows only linearly with *n* and can also be parallelized easily. Thus, it is not *substantially* critical for massive data.

The theoretical computational complexity of this approach is compared to the one of the master model calculation in Figure 5.17. As can be seen from the formula given in the lower part of the figure, the reduction of complexity is achieved by removing *n* from cubed terms (except *n mod s*, which is not critical).

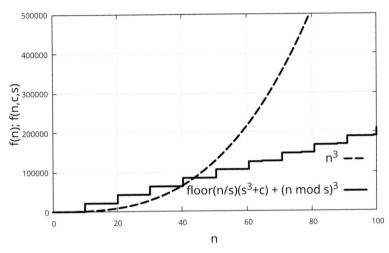

Figure 5.17: Theoretical computational complexity of master model calculation (dashed line) vs. the sequential calculation method (solid line); n = all samples, s = size of sub-model, c = merging effort.

Assuming this merging procedure, spatially isolated or temporally outdated observations can keep their influence over multiple merging steps, depending on the decay function (Equation 5.19). This is especially helpful when no better observations are available to overwrite them. Nevertheless, by using the kriging variance, the growing uncertainty of such an estimation can be expressed, which can then be considered where it appears relevant for monitoring and analysis.

Apart from some loss of accuracy, the strategy of sequencing comes along with several advantages. So it can be used to calculate large datasets with less computational effort. This can be carried out while, in principle, the advantages of kriging like the unbiased and smooth interpolation of minimum variance and the estimation of uncertainty at each position, are retained.

Given a continuous sensor data stream, this approach can integrate new measurements seamlessly into the previous model at flexible update rates. An experimental evaluation of this concept will be presented in Section 7.3.

5.4.3   Compression and Progressive Retrieval

The properties of sensor data about continuous phenomena allow for a specific approach when this data needs to be compressed. An algorithm and associated data structure supporting both lossless and loss prone compression is introduced here.

5.4.3.1   Overview

Data compression is one key aspect when managing sensor data streams. Notwithstanding the technological progress concerning transfer rate, processing power and memory size: it tends to be outperformed by the ever-growing amount of available observations [42].

The increased mobility of sensors due their miniaturization and improved energy efficiency extends their capabilities and therefore their areas of application. On the other hand, more advanced techniques of data processing and analysis are required to exploit these new opportunities. For achieving high efficiency, compression methods should take into account the specific structure of the data they are applied to.

Sensor observations typically describe continuous or quantitative variables in multiple dimensions like latitude and longitude, time, temperature, pressure, voltage, and others [109, 15]. When this data tends to be stationary in space and time, there is a high potential for compression: the actual values within a confined spatio-temporal region usually cover only a small range compared to the domain represented by the respective standard data type like *floating-point numbers*.

In order to exploit this circumstance for compression, a partitioning of observations by spatial, temporal or other criteria (or a combination of them) into data segments is carried out. The creation of such data segments is already reasonable for storage and retrieval using spatial or spatio-temporal databases.

One central feature of the proposed concept is that it supports progressive data loading for applications that do not (immediately) need the full accuracy of the queried data. This is especially useful for environments with limited transmission rate, image resolution and processing power like for mobile computing.

For this purpose, a recursive binary subdivision of the multidimensional value space is suggested. For a given level of progression, an identical accuracy (relative to the total range of values) can be achieved for each dimension. When using a database as a sink, it is reasonable to store those data segments as BLOBs (Binary Large OBjects) indexed by the dimension(s) used for partitioning.

Queries defined by (spatio-temporal) bounding boxes then have to be processed in two steps: First, the data segments affected by the query are identified. In the second step, the data segments are progressively decoded and transmitted until the required accuracy (e.g. for scientific analysis, web mapping or mobile computing) is achieved.

5.4.3.2   Related Work

There are other compression techniques in the context of sensor observations that are discussed in literature, which are introduced in the following.

Medeiros et al. [86] suggest a Huffman encoding applied to differences of consecutive measurements and thus achieve high compression ratios. This method works very efficiently with a time series of single sensors for one dimension with small changes between consecutive observations.

A more adaptive approach of Huffman encoding is introduced by Kolo et al. [67], where data sequences are partitioned into blocks which are compressed by individual schemes for better efficiency.

Sathe et al. [112] introduced various compression methods, mainly known from the signal processing literature. They are restricted to one measurement variable of one sensor.

Dang et al. [32] propose a virtual indexing to cluster measurements which are similar in value but not necessarily spatio-temporally proximate. After this rearrangement, the data is compressed using discrete cosine transformations and discrete wavelet transformations.

The compression of multidimensional signals is covered by Duarte & Baraniuk [36] and Leinonen [74]. Both works apply the Kronecker compressive sensing approach exploiting sparse approximations of signals with matrix algebra and is of high computational complexity.

Huang et al. [58] apply octree subdivision and exploit the proximity of values that often correspond with spatial proximity within octree-cells. The focus here, however, is 3D visualization with specific coding techniques for colors and meshes of different detail levels instead of multidimensional continuous fields.

5.4.3.3   Requirements

The works listed above make use of the strong correlation of consecutive sensor measurements for compression. The compression method introduced here does not presume such order. Instead, it addresses the following requirements simultaneously:

- The compressed units of data are to be organized as spatio-temporally confined segments suited for systematic archiving in spatial/spatio-temporal databases
- Diverse data types, namely *Double, Integer, DateTime* and *Boolean* can be compressed without losses
- Compression/decompression of multiple data dimensions is performed simultaneously
- Within one data segment, observations are compressed independently (no consecutive observations of single sensors tracked by their IDs are considered) and thus can handle data from mobile sensors which are arbitrarily distributed in space and time
- Data can be decoded progressively, e.g. for preview maps or applications with limited accuracy demands
- Computational cost for coding/decoding is low ($\mathcal{O}(n)$)

5.4.3.4   *Principle*

The principle that is applied for the compression method is derived from the Binary Space Partitioning tree (BSP tree, [111]). Unlike its common utilization for indexing, it is used here as a compression method that is applied to each single observation in a dataset. It does not presume high correlation of consecutive observations (time series), like e.g. Huffman encoding does [67, 86]. Consequently, the algorithm does not need to keep track of individual sensors within a set of observations, but encodes each observation individually within the value domains given per variable dimension.

The general idea behind the design is to encode observations describing a continuous phenomenon within a (spatio-temporal) region. The focus is on the representation of the continuous field as a whole, not on the time series of individual sensors. With this in mind, it appears reasonable to filter out observations that do not significantly contribute to the description of the field before long-term archiving of the data. When embedded into a monitoring system, the approach would perform best after some deliberate depletion based on spatio-temporal statistics (see Section 5.4.1 and [80]).

Progressive decompression can support different requirement profiles and is thus another important design feature of the approach. For some applications, it might be reasonable to give response time behaviour (at least for first coarse results) a higher priority than full accuracy after performing one step of transmission. The specific structure of the binary format supports this claim.

5.4.3.5   *Binary Interval Subdivision*

For each n-dimensional set of observational data, first the n-dimensional minimum-bounding box over the values is determined. In the following, the minimum and the maximum value of a dimension are denoted by *min* and *max*, respectively. The interval [*min,max*] will be called value domain. It is entailed in the domain that is covered by the corresponding data type.

Assuming the region of interest to be spatially and/or temporally confined and the phenomena observed to be of stationary character like temperature, there is a good chance for the value domain to be relatively small. Thus, a high resolution is achieved while requiring relatively few bits of data by using the multidimensional recursive binary region subdivision.

The principle is depicted for one dimension in Figure 5.18, where an interval is recursively partitioned by the binary sequence 0 − 1 − 1. The circle with double arrow represents the position within the interval of maximum possible deviation defined by that particular sequence of subdivision steps (in the following also called levels).

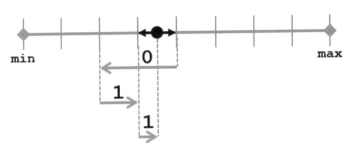

Figure 5.18: Binary space partitioning to determine a point (with maximum deviation indicated by arrows) within a value domain.

As can easily be concluded from Figure 5.18, the number of necessary bits depends on both the required absolute accuracy and on the value domain.

The considerations above provide a one-dimensional perspective on the problem. For sensor data streams, this principle has to be applied to specific data types common in this context.

5.4.3.6 *Supported Data Types*

In a sensor web environment, the collected data can in principle be of nominal scale (e.g. type of substance), ordinal scale (e.g. Beaufort wind force), interval scale (e.g. date and time) and a ratio scale (e.g. temperature in kelvin) [85, p. 16].

In the domain of data management and programming, this kind of information is usually represented by the data types *Integer*, *Float* or *Double*, *Boolean* and *DateTime*. Within a dataset or observation epoch, the actual data range is usually only a small fraction of the range covered by the respective data type.

Since the data types mentioned above have different characteristics, they will have to be considered specifically when applying the multidimensional progressive compression.

Double/Float

The compression is most straightforward for this data type. The binary tree depth n can be determined by

$$n = log_2(\frac{d}{a}) \tag{5.20}$$

where d is the value domain ($max - min$) and a is the accuracy or maximum deviation.

Within a multidimensional setting, the *relative* accuracy of each dimension is equal for equal n, while the absolute accuracy also depends on the size of its respective value domain.

Thus, when performing the compression synchronously for all dimensions with each step or level, as suggested here, equal relative accuracy for each dimension is achieved. This does not apply when one dimension has already reached its maximum bit depth, while others still have not (see Listing 5.1) or when particular dimensions have more than one bit per level to achieve faster convergence (see Listing 7.4 in Section 7.4).

In the case of a *Float/Double* data type, the interval depicted in Figure 5.18 directly represents the minimum and maximum values of the value domain, and the double arrow represents the accuracy or maximum deviation reached at that particular level (here: $0 - 1 - 1$).

Integer

Although at first glance the data type *Integer* ought to be *less* complex, it is in fact somewhat more difficult to handle with respect to progressive compression. First, the *fencepost error* has to be avoided when compressing/decompressing to the last level. So if an interval shall represent *three* integer segments, as depicted in Figure 5.19, it has to be extended to *four* segments before calculating the value domain to achieve a correct representation on the scale.

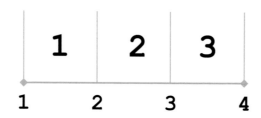

Figure 5.19: Fencepost error problem for *Integer* values.

If *Integer* numbers are used for nominal scales (e.g. for IDs), coarse indications within the value domain are maybe rather useless. For that reason it might be necessary to evaluate the complete bit depth with the first compression step. If a nominal value domain requires the maximum number of bits to be represented (see Listing 7.3 in Section 7.4), all data will have to be transmitted completely before the *Integer* value of this dimension is resolved. For more flexibility, individual bit lengths per step or level for each dimension are possible (see Section 7.4).

Boolean

Boolean values can be seen as a special case of *integers* with a range of 2. Consequently, only one step or level is needed to express one bit of information (last column in Listing 7.3, Section 7.4).

DateTime

Unlike the *Integer* type, the *DateTime* type appears much more complex than it is in handling at first glance. This is the case because it can be interpreted (and also is usually represented internally) as *ticks* (e.g. 100 nanoseconds) elapsed since some reference point in time (e.g. 01.01.2000, 00:00:00 h). This internal value (usually a 64-bit *Integer*) is provided by most libraries and can be used to handle the *DateTime* data type as normal *Integer* or *Double* for compression. Usually, time spans within a dataset of observations are tiny compared to the ones covered by the *DateTime* type, and the necessary temporal resolution is also by far lower than that of this data type. Thus, the compression rates for this particular data type are usually high.

The data types listed above usually cover the most information found in sensor data streams. Depending on the particular structure of a dataset, differing compression algorithms might provide better efficiency.

5.4.3.7 Compression Features

Based on the common principle of compression for the different data types, the specific features facilitated by that principle will be set out in the following.

Parallel compression of all Dimensions

One central feature of the proposed compression format is the progressive retrieval of sets of observations with increasing accuracy with each step or level. The general format is shown in Listing 5.1 which displays the compression format for seven dimensions of one observation.

```
i        o
dxyztvn
-------
1010111
101000
100001
010001
111011
0010 1
00 0 1
01 0 0
10 0 1
0  1
1
0
```

Listing 5.1: Binary compression format for progressive sensor data storage (column names and values to be read vertically downwards); after its name, each column contains the binary representation of the value dimension with increasing accuracy per step.

Each column entails one value dimension and each row represents one level of progressive coding/decoding. The bitstream of a particular dimension terminates at the level where its preset resolution/accuracy is reached. For the data type *Boolean* (right column: *on*) this is already the case after the first step or row.

Unlike the structure displayed in Listing 5.1 for visualization, the actual binary format does not contain blank positions, but only the data bits. Therefore, for decompression it is necessary to consider the format structure to have each bit assigned to the correct dimension.

Due to its general structure, with increasing row numbers this format tends to decrease in data volume per row and finally contributes to the accuracy of the dimensions with highest predefined resolutions only.

Flexible Bit Length per Row

Working with the structure described above can lead to a situation where a particular dimension might not be determined with a desired accuracy until the last row is reached. Most of the data might have been transmitted unnecessarily because a low accuracy would have sufficed for other dimensions. This situation might particularly be the case for IDs (first column in Listings 5.1 and 5.2) or nominal scales. It might be indispensable to receive their exact value at an early stage of the stepwise transmission.

As a solution for this problem, the bit lengths per row can be set individually for each dimension (see column *id* in Listing 5.2). Thus, the value of a dimension can converge much quicker towards its actual value with each step. In the extreme case, the exact value can already be provided with the first step of transmission (as it is *always* the case for binary values). This option can be useful when the IDs of observations are needed immediately for visualization or mapping with other data sources.

```
i               o
d   xyztvn
_____
111010111
10001000
10100001
00010001
    11011
    010 1
    0 0 1
    1 0 0
    0 0 1
    1
```

Listing 5.2: Binary format with flexible bit length per dimension; here, dimension *id* is coded with three bits per row reducing the necessary rows to four instead of twelve.

Progressive Decompression

As a consequence of the special data structure introduced here, the decompression process must permanently keep track of the actual bit configuration and the number of bits sent so far. With each new row transmitted, there is an improvement in the accuracy (the factor depending on the number of bits per row) of each dimension. In an environment with bandwidth restrictions, this progressive method provides immediate coarse results, e.g. for visualization. With the last step, the data is transmitted completely lossless according to the predefined resolution. This is not always necessarily the best choice since the data might not be needed immediately in full accuracy but rather within a shorter transmission time (responsiveness). The transmission can therefore be aborted at any level.

An experimental evaluation of the compression concept set out here is carried out with buoy data in Section 7.4.

5.5   Generic Toolset for Variation and Evaluation of System Configurations

Several approaches introduced in this chapter so far deal with the fact of limited resources in the context of monitoring scenarios. This is the case for the estimation of efficient sampling (Subsection 5.3.2), the aggregation of the experimental variogram prior to parameter adjustment (Subsection 5.3.4), the sequential interpolation (Subsection 5.4.2), and the compression of observational data (Subsection 5.4.3).

Beyond such specific improvements of efficiency, this section introduces a general framework for a continuous systematic enhancement that is driven by key performance indicators.

Variations of methods and associated parameters will affect the output of a simulation scenario. A continuous optimization of the whole process can only be carried out with appropriate output performance indicators expressing both quality and efficiency.

Following this principle, a general concept for systematic variation of methods and parameters and the inspection of their effects on output indicators is introduced. It abstracts from the particular algorithm at hand and provides a generic toolset for simulation environments.

5.5.1 Context and Abstraction

As already argued in Chapter 2, a monitoring system should be designed to provide sufficient model results with the least resources possible. The aspect of resource requirements and performance indicators in a monitoring scenario is illustrated in Figure 5.20 and specified in the following.

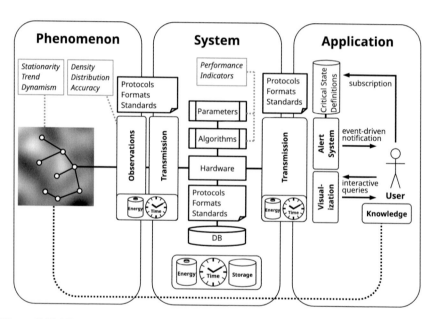

Figure 5.20: Elements of monitoring taking into consideration the limited resources, energy, time and storage.

First and foremost, the phenomenon needs to be observed and the observational data has to be transmitted to the system. The system processes and archives the data in a way that provides information of higher generality, abstraction and therefore of higher value to applications. Specifically, the improvement takes place on several levels:

- coverage
- accuracy
- density
- interoperability
- interpretability
- usability

In other words: by deploying resources for computation, transmission and storage, a monitoring system transforms raw observational data to more valuable information according to the aspects listed above. Using these resources efficiently is the objective of any monitoring system.

Applications of various kinds can make use of such higher-level services provided by the system. Details about interpolation can thus be decoupled from the application logic [119, 38]. The concept of a field data type [76, 20] is one of the key features to achieve this goal.

The overall objective is to provide knowledge about the phenomenon that is in some way useful. Since resources for such monitoring are limited (see Section 2.3), the challenge is to find a good compromise between cost and benefit.

The means to establish such a monitoring are sensors, communication networks, computers, algorithms and their associated parameters, and standards for transmission and interoperability, as depicted in Figure 5.20. The hardware-equipment of a monitoring system should be configured following the principles formulated in Section 2.4 and balancing the factors featured in Figure 2.2 (p. 34).

The effectiveness and efficiency of such a monitoring system needs to be estimated in the planning phase, but also needs to be evaluated and improved when the system is operating. The most crucial decision is about accuracy, density and distribution of observations (see Figure 5.20; also Sections 2.3.2 and 5.3.2). At this stage, the degree of knowledge about a phenomenon is determined since even the most sophisticated processing methods cannot compensate insufficient sampling.

In order to be processed for a whole region, the observational data needs to be transmitted and collected within a sensor network. Appropriate transmission protocols and data formats should be used in order to minimize time, energy and data volume.

Once the data is available in the central system, complex operations like spatio-temporal interpolation can be performed. Hardware, algorithms and the amount of data determine the expense of resources here. Variation of algorithms and adjustment of parameters can improve quality and efficiency, which will be reflected by the associated performance indicators. Persistent storage preferably is carried out on a database, supporting spatio-temporally referenced data and fostering efficient retrieval; compression (Section 5.4.3) can reduce storage space and transmission effort.

The main processing component of the monitoring system (named 'System' in Figure 5.20) is adding value to the observations in the sense that gaps are filled and the provided format is by far more interoperable than the original sensor data. In the ideal case, the phenomenon is presented as a continuous model with a good estimation of the variable of interest at arbitrary positions in space and time within the observed area (see Chapter 6).

Kriging also provides the estimation variance for each position, which can be used for smooth updates and performance improvement (see Section 5.4.2), but also as an adaptive filter (see Figure 5.13, p. 90) and to express the confidence about an aggregation statement (see Section 6).

Being presented in this derived form, the data about the phenomenon can easily be explored interactively or accessed by applications, probably via a web service. More complex services like an alert system based on aggregated data (e.g. notifying about an exceeded daily threshold for a region, see Chapter 6, Section 7.6) can be constructed when such an infrastructure is available.

The desired quality of such services determines the minimum costs and efforts necessary to establish them and keep them operational. An experimental setup as introduced in this work can significantly contribute to improve efficiency and reduce these costs.

Following the objective of balancing cost and benefit of the system, an iterative optimization is carried out to get the best possible results from limited resources. In the context of computing systems, the expenses in time, energy and storage are most relevant to be considered to achieve a result of particular quality.

Time for processing is critical when hardware power is limited, whether for financial or technical reasons. Energy is most critical for small battery-powered systems as well as for big systems like mainframe or cluster computing systems. Data volume should be reduced where possible to unburden transmission and archiving. Efficient processing can significantly reduce costs here.

This section focuses on potentials for optimization by variation of algorithms and parameter settings. To *systematically* and *reproducibly* evaluate the efficiency of each variant, a generic concept for a quantification of the following aspects is needed:

- workload
- resources
- output indicators

The workload is the computational effort that is necessary to process a particular input dataset by a particular algorithm with a particular corresponding parameter set.

The resources represent the computational hardware that is available for this task. While it is rather straightforward to quantify the resource *storage* by bits or bytes, the processing unit can be described in terms of frequency and number of kernels and other properties. Therefore, its performance might depend on whether the implementation of the algorithm supports multithreading or not. This issue is addressed in the next subsection.

The output indicators quantify the quality and other benchmarks like expenses in time and energy. These two do depend on the constellation of two other input indicators: workload and computer resources.

Algorithmic optimization and parameter tuning affects the quantity of workload and therefore also the output indicators, as Beven [13, p. 11] points out:

> For each combination of parameter values, we can calculate a model response.

Doing so while overlooking the effects that different parameter settings have on the different output indicators can then be regarded as an evolutionary process towards increasingly better solutions [44].

One important intention behind the proposed model is to quantify efficiency improvements independently from the hardware configuration that the simulation is currently calculated on. When this quantity is combined with a concrete hardware configuration, expenses in time and energy can be derived.

Especially for wireless systems, the estimation of the actual temporal and energetic expenses on a particular hardware constellation can be crucial in the planning phase. Such a transfer of processing expenses can only be carried out with a generic concept for a computational workload, which is introduced in the next section.

The automatic variation of algorithms and corresponding parameter settings is the second objective necessary for systematic evolutionary improvement. In order to handle and evaluate configurational settings of arbitrary complexity, a generic hierarchical structure is introduced in Section 5.5.3.

Together, the two components form a powerful toolset to systematically test and evaluate numerous configurations concerning hardware and algorithms in complex processing scenarios.

5.5.2 Computational Workload

When processing tasks are so complex that their execution might exceed critical resource thresholds, the resource requirement for a particular workload (or *job*, meaning "information-processing task" [39, p. 225]) is often specified by execution time on a particular machine. It is easy to determine and for many purposes provides a sufficient estimation.

This kind of metric, however, has several drawbacks because it strongly depends on the system it was actually measured in. Given the hardware specifications of this system, it might appear easy to predict the execution time for a system with different hardware. In practice, however, it can not simply be concluded that, e.g. double CPU clock speed means half execution time. Many other factors like bus frequency, amount of memory, number of processor cores, and implementation details of the program will also affect the overall performance [40].

From a practical viewpoint, it might not pay off to consider all these factors right away. Instead, a model should initially contain only the most influential factors and be equipped by additional factors only if it proves to be inadequate [72, p. 8]. With respect to this principle, the properties to be considered in this work are the CPU speed, the number of logical processors and the capability of critical code sections to run on multiple threads.

In order to predict the performance on different platforms, the central objective is to describe a particular computational workload in a way that does not strongly depend on the execution environment. The central idea to achieve this is to decouple the logical instructions from the physical resources like CPU speed [39, p. 225]. This separation makes it possible to estimate the processing time for a properly described workload without actually having to execute it on particular machines.

This can be indispensable information when a particular response time has to be granted for a monitoring service and sufficient hardware must be deployed. Such considerations can even be more important for wireless sensor networks since workload quantity is, to a certain degree, proportional to energy consumption on identical hardware.

As a consequence of the considerations above, the workload model and the execution environment or hardware system model have to be defined separately, but with associations to each other according to logical and physical properties and resources. A prototypic realization of this general concept is given by Figure 5.21.

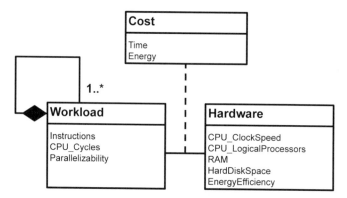

Figure 5.21: Generic structure to quantify computational cost: class *Workload* represents the machine-independent workload quantity units (as a composite aggregation [71, p. 414 ff.]) with differentiation between parallelizable and non-parallelizable sequences; class *Hardware* specifies performance-relevant properties of a processing unit; class *Cost* determines the expense in time and energy resulting from such a given association between *workload* and *hardware*.

The central feature of this structure is the systematic differentiation and aggregation of code segments according to their parallelization capability. This is of primary relevance because modern computer systems increasingly utilize parallelization [23], and, consequently, so do complex applications like spatial interpolation [100, 130, 61]. In principle, the same pattern is applicable in scenarios that work with graphics processing units (GPU) [57].

In order to obtain an abstract and machine-independent description of a particular workload, all of its logical instructions need to be counted while keeping track of their capability for parallelization. Within a complex computing task, there will usually be sequences which are implemented using parallelization, but also ones which do not, for example, because it is not possible (serial algorithms) or because the expected efficiency gain does not justify the implementation overhead.

By defining the total computational cost as a composition of sub-portions, as indicated by the UML class diagram (Figure 5.21), it is possible to divide the entire workload into portions that can be specified individually according to their parallelization capability.

Assuming the temporal effort for a particular hardware configuration as metric, these portions sum up to the entire workload by

$$t = \sum_{i=1}^{n} \frac{gc_i}{f \cdot fc \cdot thr_i \cdot thro_i}, \tag{5.21}$$

where gc_i is the computational workload of the portion i of the algorithm, expressed as unit *gigacycles* (billion processor cycles), thr_i is the number of threads this portion can be calculated with (will be *one* for non-parallelizable parts) and f is the CPU clock frequency in *GHz*. In addition, to take into account product-specific differences in the number of instructions that can be processed per cycle, the factor fc is introduced. It might either be determined experimentally or derived from product specifications. With $thro_i$, the overhead of multithreading is also considered for each portion of code. It is set to 1.0 if multithreading is not carried out.

Equation 5.21 represents the class *Cost* from Figure 5.21 by combining the machine-independent parameter gc_i with the other, machine-dependent parameters.

The quantity of *gigacycles* might be obtained or estimated in different ways, e.g. by using external performance evaluation tools. Integrating this task into the development process—i.e. into source code—provides maximum control and extensibility [114, p. 419 f.]. For that reason, this approach was chosen for the framework introduced here.

For the experimental evaluation as set out in Section 7.5, the quantity *gigacycles* is obtained by the C++ function *QueryProcessCycleTime* that is imported as external code to the C# environment. Although the term *time* within the function name indicates the physical unit, it actually provides the number of all CPU cycles of the calling process since it started. The function sums up the cycles from *all* running threads, so this is the value that is to be stored as an attribute of the *Workload* item as defined in Figure 5.21, with the *Parallelizability* attribute set to *true* if implemented accordingly.

Given a set of *Workload* objects which were deliberately registered with respect to the capability of parallelization of the respective code, it is straightforward to translate this structured quantity into processing time on a particular hardware (Equation 5.21).

While consumption of *time* is crucial for complex processing tasks and real-time monitoring applications, the consumption of *energy* is especially critical for battery-operated devices as used in wireless sensor networks.

To check the operability of such a system and to optimize it according to energy efficiency, it might be necessary to estimate the energy consumption for a particular hardware configuration. Therefore, a rough estimation of the total energy consumption w, e.g. stated in the unit *nanojoule* that is necessary for a particular process, can be given by the equation,

$$w = \sum_{i=1}^{n} gc_i \cdot w_i, \tag{5.22}$$

where w_i is the amount of energy consumed per gigacycle in each portion gc_i of the algorithm.

Individual values for each process portion can be considered where different amounts of energy per cycle are consumed, eventually depending on whether it is parallelised or not. As already mentioned, there might also be portions of an algorithm that can be delegated to a graphics processing unit (GPU) or field programmable gate array (FPGA) [78], which would eventually call for individual specification.

The aspect of energy consumption is not covered beyond this conceptual level here. However, in the context of wireless sensor web scenarios it appears reasonable to also simulate energy consumption per processing unit in order to find efficient monitoring strategies. Given the general structure as described above and as formalized in Figure 5.21, the model can easily be extended with respect to energy consumption.

The quantities given by Equations 5.21 and 5.22 can only be seen as approximations since there are many aspects which can blur such calculations. A closer consideration of the following factors might therefore be necessary when the concept is to be refined (see also [40]):

- hardware design: CPU, memory access, pipelining
- multithreading management overheads (synchronisation and others)
- processing portions delegated to a GPU
- programming language (e.g. garbage collection)
- compiler optimizations (e.g. JIT compiler effects [93])
- operating system
- different number of instructions per clock cycle

Where necessary, these blurring influence factors can be estimated and included into the equations. In summary, the concept of a machine-independent workload metric is at least a rough but systematic approximation necessary to test algorithmic variants with regard to several performance indicators for different system configurations. It is an important contribution towards iterative optimisation, especially for real-time monitoring or distributed systems like wireless sensor networks.

5.5.3 Systematic Variation of Methods, Parameters and Configurations

For a monitoring scenario as set out in this work, there is a variety of method variants, parameters and configurations that need to be evaluated with respect to their performance. One possible approach for testing different configurations is to vary *one* parameter while leaving other parameters fixed and regard the resulting series according to some evaluation metric and thus determine the best variant of that parameter [116]. Repeating this procedure for *n* parameters reveals a set of parameter configurations which might be considered appropriate for the given process. The number of variants to be tested is therefore the *sum* of variants per parameter.

But there is a fundamental problem with this approach: There has to be an initial configuration for *all* parameters for which the variation of *one* parameter per testing epoch is performed. This initial configuration has to be often chosen arbitrarily. Favourable constellations of parameters might therefore remain undetected because they are not tested in this scenario.

Alternatively, *all* possible constellations of parameter settings can be considered. The total number of tests to be executed is then the *product* of the number of variants per parameter instead of their *sum*. This might of course place a considerable burden on testing scenarios.

For example, by varying only ten parameters by only ten values or options each will result in 100 configurations to check for [64, p. 61]. However, it systematizes the process and makes it far less arbitrary. The results generated by such a systematic survey of variants allows for more systematic and extensive analyses and therefore promote a deeper understanding of the whole process that is evaluated.

In complex systems like spatio-temporal analysis tools, monitoring environments or simulation frameworks, algorithmic variants and associated variable parameters tend to increase over time. As a consequence, the manual or semi-automated setting and evaluation of variants, e.g. by configuration files, become increasingly cumbersome, error prone and arbitrary. A generic software solution for this problem is sketched as a UML class diagram in Figure 5.22.

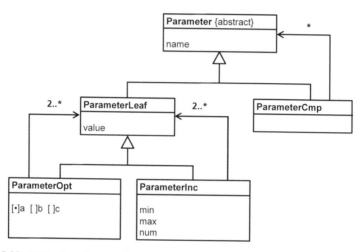

Figure 5.22: UML class diagram for generic organisation of configuration variants based on the composite pattern [71]: class *Parameter* as abstract concept and *ParameterCmp* as container for structured organization; class *ParameterLeaf* representing the instance actually containing the value, concretized as option (class *ParamterOpt*) or as increments within an interval (class *ParameterInc*).

The composite pattern supports hierarchical organisation and polymorphic treatment of whole-part-relationships of objects [71, 118]. We exploit this capability of the pattern here to define a generic structure which can handle arbitrary complex configurations of parameter variants in a uniform way. The class *Parameter* represents an abstract concept that can be instantiated as a list of mutually exclusive options (class *ParameterOpt*), an interval division comprising integer or float type increments (*ParameterInc*), the corresponding enumeration or numerical value itself (*ParameterLeaf*) or a named composite of multiple parameters (*ParameterCmp*).

This pattern is chosen for its capability to reflect the complex hierarchical parameter structure which is present in many systems. By using recursive polymorphic function calls, all possible configuration variants can be created and iterated [118, p. 83 ff.], [87, p. 227 ff., p. 225 ff.].

With this structure it is also possible to organize parameters in sub-trees that describe logical units within the modelled environment. So a hardware configuration as described in Section 5.5.2 can be subsumed as a composite type *ParameterCmp* containing parameters for a number of CPU cores, CPU clock speeds and RAMs. Variation can then be carried out by each parameter (e.g. 1, 2, 4, 8 CPU cores; 1, 2, 4 GiB RAM, 1.2, 2.0, 2.6 GHz CPU clock speed) or, alternatively, by predefined named configuration sets (e.g. Raspberry Pi® 3 Model B, Dell Precision® Tower 5810) and the corresponding detailed information parameter set. The variation can then take place by switching between those named subsets.

With this structure, variations of preset hardware configurations can be carried out just as easily as any other set of parameters within a simulation. An estimation of processing time for machine-independent workloads can then be carried out for each of those configurations. This functionality can be important while considering the hardware equipment for a planned monitoring system.

There might also be constellations where environmental conditions that affect data transmission—like meteorological parameters—should take part in the systematic variation. Analogous to hardware configurations, such constellations can also be modelled and handled as named composite subsets.

The guiding idea of the structure described above is to automatically generate the list of all possible configurational variants instead of having to update such a list manually with every added or removed parameter option. In order to generate a series of simulation variants based on this structure, it has to be integrated into the superordinate flow control of the simulation framework.

For systematic evaluation of those variants, their output performance indicators are stored in tabular form using the naming conventions of the identifiers for algorithms and parameters (see Tables 7.2 and 7.5 in Section 7). The same conventions are used for the directory structure the output data is stored into. The indicator log file relates configurational variants to performance indicators for each process run and thus facilitates deeper and more systematic analysis.

5.5.4   *Overall Evaluation Concept*

The main objective behind the concept proposed so far is to organize variants of analyses or simulations and evaluate the different outcomes with one or more performance indicators. Which input properties and performance indicators are to be taken into account for such experimental series depends on the particular application.

Table 5.1 provides a representation of the general concept of such a relation. For a wider perspective, it is extended by properties and indicators that can be considered reasonable in the context of environmental monitoring.

Input Properties contains all the items and characteristics which constitute the environment and the monitoring system as a whole: the environmental conditions that influence monitoring, hardware, methods, tools, datasets and formats are used to generate the model. For a given scenario of input properties, the

Table 5.1: Input properties (arranged by the categories environment, hardware, data and algorithm) and output indicators of complex computing systems; their interdependencies are indicated by dots.

Input Properties		Time	Energy	Data Volume (MB)	Transmission Effort	Accuracy	Compression (Rate, Loss)
Env.	Phenomenon Dynamics					•	
	Sampl. Rate/Distr.		•	•	•	•	
	Transmission Medium	•	•		•		
Hardware	Sensor Energy Efficiency		•				
	Comm. Bandwidth	•			•		
	Comm. Energy Efficiency		•		•		
	CPU clock speed	•	•				
	CPU logical processors	•	•				
	RAM	•	•				
	Storage	•	•				
	Computational Efficiency		•				
Data	Input Data (Amount/Format)	•	•	•	•		
	Data Density (Raster/Vector)	•	•	•	•	•	
	Compressibility	•	•	•	•		•
Algorithm	Method Set	•	•	•	•	•	•
	Parameter Set	•	•	•	•	•	•
	Parallelization Capability	•	•				
	Compression Method	•	•	•	•	•	•
	Indexing Method	•	•	•			

Output Indicators represent the metrics that can be used to evaluate the whole process chain. The interdependencies between these items are specified in the following.

Environment

The observed phenomenon can be described by its dynamism in space and time. Leaving all other elements of the monitoring system unchanged, it only affects the accuracy of the model. If the phenomenon has more dynamism than is covered by the sampling layout, it will affect the model accuracy.

Changing the sampling rate affects the effort that is necessary for sensing, computation, transmission and storage, but it also changes the output accuracy. The transmission medium entails atmosphere, topography and also potential sources of interference that affect the transmission which can also be associated with energy. If the transmission signal is deceasing [3] and the message has to be repeated due to errors indicated by the protocol, temporal delay may also occur.

Hardware

The properties of the hardware involved in monitoring are listed in the next group of Table 5.1. Leaving factors like workload or algorithm unchanged, the energy demand is affected by the efficiency of sensors, communication devices and processing units. The throughput per time unit depends on hardware specifications and communication bandwidth while the effort for transmission according to time and energy depends on the efficiency and bandwidth of the communication devices.

When plenty of hard disk storage is available, time and energy performance can be increased by providing multiple indexes, controlled redundancy and preprocessed data (e.g. discretization of continuous fields by raster grids, see Section 2.2.3 and Chapter 8).

Data

From the data perspective, it is the amount and format of incoming sensor observations that affects most of the output indicators. The data *density* in the context of the *Data*-group in Table 5.1 does not refer to the raw observational data, but rather addresses archiving and retrieval. Vector data can be thinned out deliberately while minimizing information loss under the presumed interpolation method (see Figure 5.13). Raster grids can be provided statically or dynamically in different resolutions depending on the requirements.

Having efficient transmission and storage of data in mind, the compressibility is another important factor. The structure of observational data of continuous phenomena allows for good compression rates and progressive retrieval (see Section 5.4.3). This reduces the data volume, but goes along with extra effort for compression and decompression, which might affect response time. This also might affect energy resources, since compression/decompression is less expensive than data transmission [6]. The compression rate depends on both the method and the data. Sophisticated algorithms prone to losses for raster grids achieve high compression rates with small accuracy losses [104].

Algorithm

In many cases, the choice and configuration of *Algorithms* is the most obvious way to influence the output indicators. The set of methods with their associated parameters can have an impact on any indicator. It is also significant how the particular algorithm was implemented. The parallelization capability depends on both the implementation and the hardware. It primarily affects processing time and hence the response time as a result. Since it is entangled with data volume and accuracy, the compression method can affect all indicators. Indexing reduces search operations and therefore saves time and energy while increasing necessary data volume.

As has been shown above, the interdependencies between input properties and output indicators within a monitoring system are manifold. For real world scenarios, they might be more complex than is indicated by Table 5.1. A systematic inventory like this, however, does support the process of decision-making when establishing or auditing an environmental monitoring system. Beyond the accuracy-centred model evaluation as suggested by Beven [13, p. 3], it also takes environmental, technical and resource-based matters into consideration. Thus, it can help to systematically evolve a monitoring system towards better quality and efficiency.

5.6   Summary

In this chapter, various methods and algorithms have been set out that are necessary for an efficient monitoring of continuous environmental phenomena. Besides the central task of interpolation, also the generation of continuous random fields, their sampling and the merging of sub-models (for efficiency and smooth differential updates) have been addressed.

The proposed features constitute a powerful simulation environment that is capable of testing manifold modes and configurations of environmental monitoring. In order to facilitate this task of carrying out and evaluating experiments with multiple configurations, a generic software solution has been proposed. It automates configurational permutations and relates parameter settings to output datasets and associated indicators in order to systematize the evaluation process.

The simulation experiments which are introduced in Chapter 7 have been carried out to evaluate some of the crucial methods that were introduced here. The tools for systematic variation and evaluation have been extensively used in these experiments.

Chapter 6

A General Concept for Higher Level Queries about Continuous Phenomena

CONTENTS

6.1   Introduction 120
6.2   Interpolation 121
6.3   Intersection 124
6.4   Aggregation 125
6.5   Conclusions 127

6.1 Introduction

Environmental monitoring is carried out by society for reasons of protection, profit, scientific progress among others (see list on page 28). By using sensors, it transmits a particular environmental state to a system that is based on an abstract model (see Figure 2.1 on page 23). The quality of the observations, the model, and the system have to comply with the requirements as stipulated by a particular task or question.

In the context addressed here, we deal with a model about the particular phenomenon (e.g. temperature) which behaves *continuously* in space and time. Small changes in space and/or time within the covered space-time-cube will usually show only small changes in its value. The continuum is either determined by differential equations which are applied to starting conditions (as with fluid dynamics) or as optimal estimations filling the gaps between given observations by interpolation (as with statistics for spatio-temporal data). Mixed approaches can be applied in order to improve the quality of an estimation [2, p. 313].

In any case, the set of given observations is used to estimate the parameters of the given model and to adapt it to the particular phenomenon. The model is then capable of estimating the values of interest between the observations. As a metaphor, we can imagine a virtual sensor that provides the particular value at arbitrary positions in space-time within the region of interest. When applying the metaphor to multiple observations, e.g. in a spatio-temporal grid, we can easily derive aggregated information like the average temperature of a certain district for a particular period of time. The grid resolution has to be chosen case by case as an appropriate compromise between accuracy and computational effort (see Chapter 4).

The virtual sensor metaphor can be seen as the mediator between the available discrete observations and the knowledge that is actually required. It is inspired by the abstraction approach associated with the concept of a field data type [76, 25].

Taken further, it can be thought of as a specialized machine that processes all available information (discrete observations) about a phenomenon according to the best scientific methodological knowledge (interpolation algorithms and parameters) in order to provide derived information at arbitrary positions in space and time (virtual sensor) while most efficiently exploiting the available resources.

As the most basic elements of the monitoring process, the performed observations, the applied interpolation methods and their parameters do fundamentally differ in kind, quality, quantity, validity and computational workload. The metaphor of a virtual sensor, though, can abstract from the particularities of the manifold methods that can be applied. It constitutes an abstract framework that can integrate any interpolation method and can control the accuracy and computational effort by varying the applied resolution. It is flexible with respect to the amount and distribution of observations.

It is capable of providing performance indicators like overall interpolation quality, computational effort or responsiveness for any interpolation method that implements its basic functionality. It can thus serve as a tool to compare different interpolation approaches according to different generic evaluation criteria.

When the statistical method of kriging is applied, it provides an estimation of the variance of the interpolation besides the interpolation itself. This information can be integrated as an extra band of the raster format (see Section 4.3). Thus, it is capable of analyses of higher abstraction levels like hypothesis testing (see Section 7.6). It can also be applied to address tasks like filtering and sequential integration of new observations to existing models (see Section 5.4).

The general approach of raster-vector interoperability as a means to handle continuous phenomena have already been sketched in Chapter 4. This chapter presumes some ideas introduced in the last chapter and applies them to possible monitoring scenarios. The aim of the concepts introduced here is to avoid ad hoc analyses in favour of queries which are expressed rather descriptively than imperatively.

6.2 Interpolation

Generally, the interpolation has the function of filling the gaps between observations. The quality of the resulting model depends on the complexity of the model, the density and distribution of observations and the applied method and its parameters.

As mentioned above, the interpolation can be interpreted as a virtual sensor that can be placed arbitrarily within the covered area. In contrast to a real sensor, the virtual sensor should provide a variance since the value is not directly measured but (statistically or deterministically) derived from the given set of observations. Besides single positions, the virtual sensor can also be used to provide interpolated values along a (spatio-temporal) trajectory.

This option might be useful where an accumulation of a variable is needed, e.g. an individual's exposure to radiation resulting from a transect taken through a contaminated area that is monitored by sensors. It might also be used for different scenarios within a simulation in order to determine an optimal route for accomplishing a mission. The general concept of a virtual sensor is depicted in Figure 6.1.

The virtual sensor is a metaphor to guide the system development process. It encapsulates the fundamental approach of interpolation as one handy concept. The crucial control properties like methods and parameters can be exchanged without affecting the general nature of the concept of a virtual sensor. It is therefore predestined as a framework to make different interpolation approaches comparable.

The universality of the concept is also applicable with respect to the type of output. By its capability to estimate values (and associated variances) at arbitrary

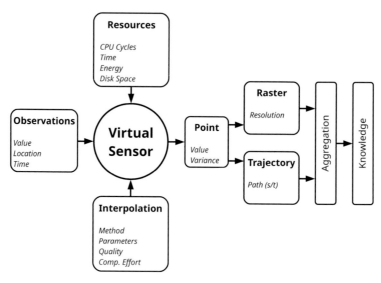

Figure 6.1: Virtual Sensor as an abstract concept for generating knowledge from observations. The virtual sensor can estimate values for arbitrary positions from a set of discrete observations. It does so by using interpolation methods and associated parameters, which provide different qualities for different computational efforts. The performance of this computationally intensive process is constrained by the available resources like CPU cycles, time, energy, and disk space. The basic output is a point with an estimated value and, in case of kriging, a variance estimation for that value. By generating multiple output points, a raster of desired resolution or a trajectory (a path, eventually with timestamp per point) can easily be derived. Aggregations applied to rasters or trajectories finally produce the required knowledge.

positions in space and time, it can produce output types like rasters, transects or trajectories simply by applying it for a particular set of spatio-temporal positions. Transects can be seen here as a specialization of trajectories having a specific geometry and identical timestamps in all elements.

Wherever performance is a critical factor, the trade-off between the quality and the computational effort of the interpolation method must be considered carefully in order to produce the required performance with the given resources. While for some applications coarse interpolations might suffice, scientific analyses will probably require the best estimations available.

In an actual monitoring system providing the processing as described above as a service, there should be mechanisms that handle the generated data intelligently and which are transparent to the user. While the general concept of the virtual sensor is to calculate interpolated values from observations on demand, the high computational workload for interpolating a raster should eventually be "conserved" for future queries. The consumer of the service does not have to keep track of the interpolation datasets, since it is completely managed by a query handler. The general architecture is sketched in Figure 6.2.

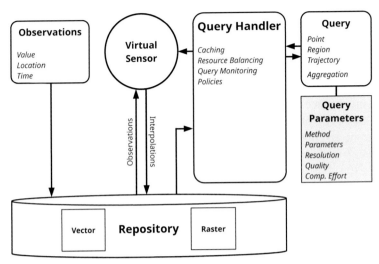

Figure 6.2: Query handler in the context of a monitoring system. A query about a continuous phenomenon entails the point, region or trajectory of interest and eventually the type of aggregation to be used. Additional properties determine the interpolation method and its parameters, the raster resolution to be used, the required interpolation quality and eventually some threshold for the computational effort. Thus specified, the query handler and the virtual sensor govern the processing that is necessary to provide a sufficient response. It caches raster data that is often queried while balancing the available memory versus the computational effort. A continuous query monitoring allows for adaptation to the requirements according to the defined policies. The repository contains both the vector data (observations, regions of interest, trajectories) and the raster data (interpolations) generated by the virtual sensor.

The main objective of this architecture is to decouple the query about a continuous phenomenon from the actual interpolation process. Whether the interpolation has to be performed on demand, produced from the cache or derived from some former interpolation raster is transparent to the client. So maybe some coarse resolution can be generated from an accurate interpolation from the repository just by resampling.

If configured by the application policy, extensive regions of the monitoring area can be interpolated while the system is otherwise idle. This might occupy much disk space, but will eventually produce a much better responsiveness in the long run. Interpolation rasters in regions not queried for a long time can be deleted again to use disk space efficiently.

In general, the decoupling of the descriptive query formulation from the actual processing allows for continuous optimization of the system with respect to its responsiveness. The client does not have to change its interface towards the monitoring system even if significant changes with respect to algorithms and resource balancing occur [119].

From a software engineering perspective, not only the query handler, but also the virtual sensor serves as an interface definition. Regardless of the particular implementation of a method and the target output format (point, raster, trajectory), the signature of the function call remains the same.

The abstraction concept introduced here is crucial according to a sound overall system architecture. The client of a monitoring service should be spared of interpolation complexity and implementation details if possible. If appropriately applied, the kriging variance can serve as a valuable indicator for the reliability of a statement about a continuous phenomenon (see Section 7.6).

Once the observations are distributed by interpolation in the region of interest by means of a raster of appropriate resolution (see Chapter 4), this raster can be used for visualization, and overlayed with other data or regional aggregation. The latter procedure presumes the grid cells involved to be identified by intersections with the region of interest, usually given by a vector geometry for the spatial dimensions and an interval for the temporal dimension. This will be covered in the next section.

6.3 Intersection

In order to identify the cells that are relevant for the analysis, the raster grid has to be intersected with the region of interest about which a statement is to be made before the aggregation can be processed (see Figure 6.3).

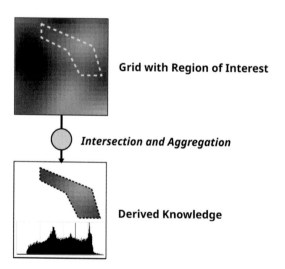

Figure 6.3: Intersecting polygons with rasters: The knowledge about a continuous phenomenon within a particular region is derived in two steps; (1) The relevant grid cells are identified by intersection with the polygon and (2) the aggregation is carried out on this subset of grid cells.

The extent of the intersection task that is necessary to identify the affected raster cells depends on the resolution that was chosen for interpolation. It represents the granularity by which the continuum has been discretised in order to gain surface-like information about it [79, p. 160]. If a sufficient raster resolution is chosen, there should be no significant difference in aggregated values for slightly differing resolutions. Once identified, the chosen aggregation method can be applied to the subset (see Figure 6.3).

6.4 Aggregation

As described above, the intention of interpolation between the discrete observations is to provide a sufficiently dense grid of values for visualization and analysis. This coverage by values of the particular variable facilitates the extraction of specific knowledge, such as the average value in a particular area within a particular period of time. The task can be specified as the aggregation of a subset of the given spatio-temporal grid of interpolated values. The (spatio-temporal) region of interest defines the borders by which the grid cells taking part in the aggregation are to be determined. The region of interest might be static as a city district within a particular time interval. It might as well be of dynamic nature such as a vehicle transect.

In more complex scenarios, it might also be derived from other continuous phenomena, e.g. when the average atmospheric pressure in regions below a particular temperature threshold is to be determined. In principle, arbitrary criteria can be applied to determine affected grid cells from an entire set.

Once the subset of relevant grid cells is found, the aggregation itself is a rather simple task and well known from classical statistics. The following aggregation might be considered reasonable in this context:

- mean
- variance
- standard deviation
- skewness
- kurtosis

Beside these standard parameters, the following might also be of interest:

- minimum value
- maximum value
- median
- mode
- mean absolute deviation (MAD)
- standard error of the mean (SEM)
- various quantiles of density

The parameters above do not take into account the specific structure of geo-statistical parameters. They assume the observations to be independent and identically distributed (IID) [46, 31], which is not the case since they are spatio-temporally correlated. In reality, they rather represent a regionalized variable with a constant mean or trend, a random but spatio-temporally correlated component and an uncorrelated random noise [19, p. 172], [21, p. 11].

A mean value in the context of geostatistics should not be confused with the best available estimation of the "true value" as known from other disciplines, where, say, a physical constant is to be determined by multiple observations. For a regionalized variable there is no "true value" since its actual manifestation as a surface spreads over a whole range where values are autocorrelated according to their distance.

The statistical properties of such a regionalized variable are therefore more complex and have to take into account the correlation of observations. This correlation structure is expressed through the variogram model. The most fundamental parameters of such a model are listed below.

■ variogram type
■ sill
■ range
■ nugget effect
■ anisotropy

The *variogram type* generally expresses how the degree of correlation depends on the spatial, temporal, or spatio-temporal distance of two observations. The parameter *sill* represents the overall variance of the set of observations regardless of their correlation. The *range* is the distance up to which observations are considered in some way spatio-temporally correlated while the *nugget effect* represents the uncorrelated random noise.

An *anisotropy* is considered if the degree of correlation does not only depend on the distance between two observations, but also on the direction of the vector connecting them. For a more thorough coverage of the variogram model see Section 3.3.

The statistical parameters introduced so far are applicable in real-world monitoring scenarios. In simulation scenarios as extensively processed in the context of this work (see Chapter 7), there are additional parameters that play an important role:

■ deviation grid related to reference model
■ root mean squared error (RMSE)

While the former represents the deviation from the reference model in each grid cell, the latter is its statistical aggregation and expresses the overall interpolation quality within a region.

In the context of an experimental setting, it is a very useful indicator when comparing the quality of different interpolation methods and parameter settings. It is the central target variable that can be automatically evaluated in simulation series with automatic variation of methods and parameters.

The experiments as they are outlined in 7.6 show a high correlation between the RMSE and the accumulated kriging variance, named root mean kriging deviation. The former describes the actual discrepancy between the synthetic reference model and the one derived from observation and interpolation; the latter represents an estimation of the overall model accuracy *without* using the knowledge of the reference model. It is derived only from the statistical properties of the observed phenomenon as given by the variogram, and the distribution of observations within the observed area.

While the correlation of the two indicators can only be shown in an experimental environment, it demonstrates the significance of confidence when performing higher level queries about continuous phenomena.

6.5   Conclusions

The fundamental goal of monitoring is to gain some abstracted portion of knowledge about specific environmental conditions. In the case of continuous phenomena, the process chain to gain this knowledge entails observation, interpolation, intersection, aggregation and also interpretation.

Within this process chain there are unlimited degrees of freedom of how to accomplish the goal. The challenge here is to specify a monitoring scenario that is sufficient for the given objective and efficient according to computational effort at the same time.

The distribution of information by interpolation and its summarization by different methods of aggregation are the core principles that transform raw sensor observations to actually valuable knowledge. A transition of information from vector to raster and vice versa (see Section 4.4) is the crucial technique by which a monitoring system delivers high level knowledge about an observed phenomenon.

While being substantial on an operational level, the details of the particular algorithms which are executed to generate this knowledge should be hidden from any higher level user. However, a measure of confidence like the kriging variance can serve as an important indicator for decision making.

It is the responsibility of the operator of the monitoring system to decide which degree of representativeness is considered sufficient for a particular phenomenon and a particular task or question associated with it. From the degree of representativeness follows the necessary effort for sensing and processing and from that the costs for establishing and operating the monitoring system (see Figure 2.2 on page 34).

Future development will provide technical environments where physical resources like sensors, networks, processing units, and storage space can easily be related with intangible items like specifications, algorithms, parameters, quality indicators, query languages, and services. Only an abstraction towards the most fundamental concepts can help to achieve a high level of interoperability and comparability of different approaches and arrangements. Some basic ideas that are following this objective have been introduced here.

Chapter 7

Experimental Evaluation

CONTENTS

7.1	Minimum Sampling Density Estimator		131
	7.1.1	Experimental Setup	131
	7.1.2	Results	131
	7.1.3	Conclusions	135
7.2	Variogram Fitting		135
	7.2.1	Experimental Setup	136
	7.2.2	Results	139
	7.2.3	Conclusions	141
7.3	Sequential Merging		141
	7.3.1	Experimental Setup	142
	7.3.2	Results	142
	7.3.3	Conclusions	144
7.4	Compression		145
	7.4.1	Experimental Setup	145
	7.4.2	Results	148
	7.4.3	Conclusions	150
7.5	Prediction of Computational Effort		151
	7.5.1	Experimental Setup	151
	7.5.2	Results	152
	7.5.3	Conclusions	152

7.6   Higher Level Queries 153
        7.6.1   Experimental Setup 153
        7.6.2   Results 157
        7.6.3   Conclusions 159
7.7   Case Study: Satellite Temperature Data 162
        7.7.1   Experimental Setup 163
        7.7.2   Results 165
        7.7.3   Conclusions 167

7.1 Minimum Sampling Density Estimator

This section provides an outline of the experiments which were carried out in order to evaluate the formula for the minimum sampling density as it was deduced in Section 5.3.2. A simulated random sampling is therefore carried out on synthetic fields. By varying the number of samples around the deduced optimum, its validity is inspected by comparing the RMSE of each interpolated model.

7.1.1 Experimental Setup

To check the validity of the approximation of the necessary minimum sampling density, the method of kriging is applied to sets of observations, varying in number, and performed on different synthetic continuous fields. Within one set, the observations are randomly and uniformly dispersed over the n-dimensional region of interest.

The differences (RMSE) between the synthetic reference field and the one derived from the interpolation are compared. The experiment is carried out on different kinds of random fields, which is specified before the respective results will be presented.

7.1.2 Results

As first reference, a two-dimensional field is generated by

$$f(x,y) = sin(x) \cdot sin(y). \tag{7.1}$$

The resulting raster grid of 150x150 pixels in greyscale levels is depicted in Figure 7.1.

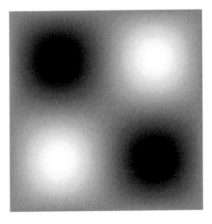

Figure 7.1: Two-dimensional sine signal as raster grid.

With Equations 5.3, 5.8 and the extent of 1λ in each spatial dimension we get

$$2\frac{1}{2^{\frac{1}{4}}} \cdot 2\frac{1}{2^{\frac{1}{4}}} = 64 \tag{7.2}$$

as an approximate minimum number of samples necessary to capture the pattern for Kriging. We take seven sampling sets from 25 up to 115 observations, increasing by 15 observations with each step, normalizing it to the calculated value of 64 and plotting this quotient against the RMSE between the reference and the derived model. For convenience, this value is normalized to the highest one in the series. The parameter *range* is also added to the diagrams. It is derived from the variogram fitting procedure (see Section 5.3.5) and is normalized to the theoretical value as determined by Equation 5.3.

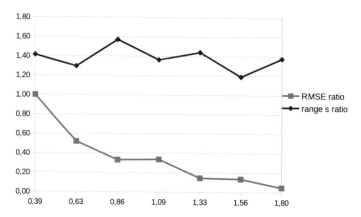

Figure 7.2: Sampling variations applied to a two-dimensional sine signal with the ratio of sampling normed to the derived value on the abscissa, and the ratio of RMSE and *range* normalized to the initial value (RMSE) and to the value of the generated field (*range s*).

As can be seen from the RMSE graph of Figure 7.2, a noticeable degree of saturation is achieved when the quotient approaches the value of one, which represents the minimum number of samples of 64 as computed by Equation 5.8.

Extending the sine signal by a third dimension reveals a similar pattern, as can be seen in Figure 7.3. In this case, the number of samples normalized in each epoch is,

$$2\frac{1}{2^{\frac{1}{4}}} \cdot 2\frac{1}{2^{\frac{1}{4}}} \cdot 2\frac{1}{2^{\frac{1}{4}}} = 512. \tag{7.3}$$

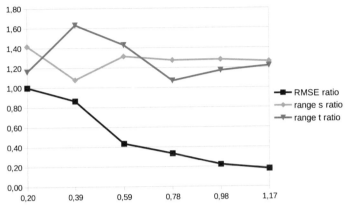

Figure 7.3: Sampling variations applied to a three-dimensional sine signal with the ratio of sampling normed to the derived value on the abscissa, and the ratio of RMSE and *range* normed to the initial value (RMSE) and to the values of the generated reference field (*range s, range t*).

Having used the separable variogram model for interpolation, the parameter *range* is separately estimated for the temporal dimension. Other models might also be applied here (see Equations 3.8, 3.9, 3.10, p. 43), but this is out of the scope of this evaluation. For the spatial dimension we assume this parameter to be equal for each direction in each experiment; otherwise anisotropy would have to be introduced [129].

The sampling of sine signals was primarily carried out for the reason of the transfer of concept of the Nyquist-Shannon theorem from signal processing to geostatistics (see Section 2.3.2). After the validity for periodic signals was shown, it was applied to continuous random fields as depicted in Figure 7.4.

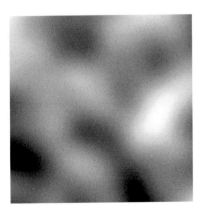

Figure 7.4: Two-dimensional synthetic random field generated by a Gaussian covariance function.

Given an extent of 150 and a range of 30, generated by a Gaussian covariance function (see Section 5.3.1), the number of necessary observations is calculated by

$$2\frac{150}{30} \cdot 2\frac{150}{30} = 100. \tag{7.4}$$

In the diagram (Figure 7.5), the effect of a saturated error quotient can again be found near the abscissa value of 1.0 that corresponds with the estimated minimum sample size.

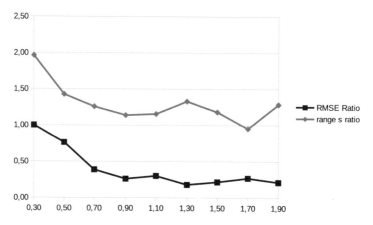

Figure 7.5: Sampling variations applied to a two-dimensional random field with the ratio of sampling normed to the derived value on the abscissa, and the ratio of RMSE and *range* normed to the initial value (RMSE) and to the value of the generator (*range s*).

In this case, the similarity between the RMSE curve and the range *s* ratio curve is striking, indicating that the accuracy of the estimation of the parameter *range* corresponds with the accuracy of the whole derived model.

This effect is less obvious in Figure 7.6, which represents sampling epochs performed on a three-dimensional random field. There is also a generally higher ratio between estimated and actual range parameter here indicating an increased uncertainty of estimation due to the higher complexity of the phenomenon. The saturation effect of the RMSE when the sample size approaches the number estimated by Equation 5.8 can nevertheless also be identified quite clearly.

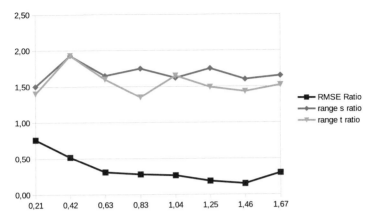

Figure 7.6: Sampling variations applied to a three-dimensional random field with the ratio of sampling normed to the derived value on the abscissa, and the ratios of RMSE and *range s, range t* normed to the initial value (RMSE) and to the values of the generator.

7.1.3   Conclusions

The experiments have corroborated the overall validity of the formula for minimum sampling density. It can thus be used to estimate the observational effort for any setting where the central geostatistical parameter *range* is known for all involved dimensions. It assumes a uniform random distribution of sample positions and can therefore only provide an approximate estimation. For the simulated monitoring scenarios it is of great value since it relieves the sampling process of arbitrariness and makes experiments on models of differing dynamism (and therefore differing values of *range*) comparable by norming the determination of necessary samples. We will make use of this formula in the subsequent experiments to reduce effects resulting from insufficient sampling or oversampling.

7.2   Variogram Fitting

Geostatistical interpolation is carried out by applying a particular covariance function with its associated parameters to generate a covariance vector for each position to be estimated. This vector expresses the correlation to each single observation by using its (n-dimensional) distance to the interpolation point as the input variable for the covariance function. Based on this structure, a linear regression is performed to find optimal weights for each observation [8, 129, 95].

In order to adapt to a particular set of observations, the parameters of the variogram model are adjusted to fit the corresponding experimental variogram (variogram fitting, Section 5.3.5). The appropriateness of the variogram model and its associated parameters determines the quality of the interpolation and is therefore decisive for the whole process.

In an experimental setting with synthetic continuous random fields as given here, the quality of interpolation and therefore the quality of variogram fitting can be expressed as the RMSE between the reference model and the one derived from interpolated observations. This is the setting that is applied in this section to identify suitable configurations for variogram fitting.

7.2.1 Experimental Setup

In order to experimentally identify favourable variants of variogram fitting, the methodological alternatives of experimental variogram aggregation (see Section 5.3.4) and fitting of the function parameters (see Section 5.3.5) are systematically tested.

Table 7.1: Methodological options for critical steps within the variogram fitting procedure.

Process	Abbr.	Parameters/Variants
Split Dimension Selection	split_dim	toggle, max_rel_dev, max_rel_ext
Split Position Selection	split_pos	mea, med, mid
Aggregation Position Selection	aggr_pos	mea, med, mid
Gauss-Newton Weighting Function	wgt_fnc	equ, lin, sin, log

The methodological parameters that are considered for systematic variation and testing are listed in Table 7.1.

To apply and compare these variants in a simulation, a continuous random field is used. It is generated by applying a moving average filter on a field of pure white noise, as described in 5.3.1. The following properties have been applied to the generating process:

- grid size of 150 x 150 (spatial dimensions) x 30 (temporal dimension) elements (=675,000 grid cells)
- spatio-temporal extent in 150 m x 150 m x 60 min
- white noise field with mean of 5000 and deviation of 500
- filter: separable covariance function based on gaussian function for spatial and temporal dimensions, spatial range of 50 grid cells ($\hat{=}$ 50 m), temporal range of 15 grid cells ($\hat{=}$ 30 min)

By transforming to greyscale levels, we get a visual impression of the model in Figure 7.7.

According to Equation 5.8, we calculate

$$2\frac{150}{50} \cdot 2\frac{150}{50} \cdot 2\frac{60}{30} = 144 \tag{7.5}$$

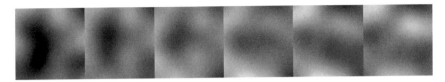

Figure 7.7: Experimental continuous random field as image sequence; images No 1, 4, 7, 10, 13, 16 out of the 30 image time series.

as the approximate number of observations necessary to capture the phenomenon adequately. The samples are dispersed randomly and uniformly over the set of 675,000 grid cells. For the experiments to follow, we keep the model and the random sampling positions constant to achieve identical conditions for all methodological variants.

By generating the experimental variogram, with Equation 5.10 we get 23,220 pairings of observations which can be investigated in terms of a correlation pattern between spatio-temporal distance and squared halved differences of values (semivariance γ, see Equation 3.2).

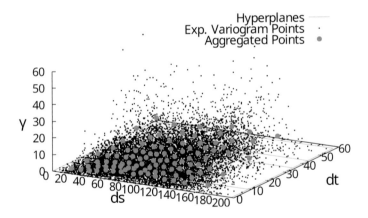

Figure 7.8: Variogram point cloud aggregation with semivariance γ for spatial (ds) and temporal (dt) distances; the hyperplanes are mainly concealed here and can be better seen in Figure 7.9.

Figure 7.8 shows the semivariance γ for each pair on the vertical axis plotted against the spatial and temporal distance on the two horizontal axes, respectively. The aggregated green points are used to fit the theoretical variogram, as can be seen in Figure 7.9. The intersection lines of the partitioning BSP planes with the plane through $\gamma = 0$ are also plotted to illustrate the prior aggregation areas.

Due to the scale of the vertical axis in Figure 7.9, the "outliers" in the regions of high spatio-temporal distance become visible. In variants where the weighting

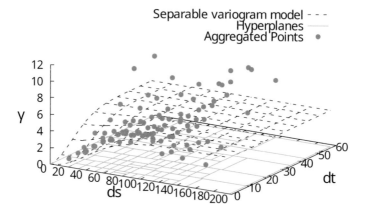

Figure 7.9: Separable variogram model fitted to aggregated points from experimental variogram point cloud with semivariance γ for spatial (ds) and temporal (dt) distances.

functions specified in Section 5.3.5 are applied, these points do not have much influence on the fitting.

To track down the particularly appropriate configurations from the methodological variants generally reasonable (see Table 7.1), all possible combinations of variants have to be processed iteratively. The number of combinations is simply the product of variants per process step by

$$3 \cdot 3 \cdot 3 \cdot 4 = 108. \tag{7.6}$$

For each of these distinct parameter configurations, the RMSE is obtained by comparing the continuous random field with the interpolation result received by this particular configurational variant.

Since the estimation of the *range* values (spatial and temporal) is the crucial step within the whole process chain, these values are included in the evaluation scheme. More precisely, the sum of relative deviations of the estimated ranges (r_s, r_t) from the ones used for the random reference field (r_{sr}, r_{tr}) are used as a metric for the quality of the overall variogram estimation by

$$d_r = \frac{\sqrt{r_s^2 + r_t^2}}{\sqrt{r_{sr}^2 + r_{tr}^2}}, \tag{7.7}$$

where small values for d_r indicate estimations of ranges near the ones used for random field generation by variogram filters (see Section 5.3.1).

To systematically compare all of the 108 parametric variants, the entire monitoring process chain is performed for each one of them.

7.2.2 Results

The results of the experiments are presented as diagram in Figure 7.10. For the best 15 variants, a more detailed view is given in Table 7.2. It is sorted by descending values of RMSE. The first four columns represent the parametric options with its selected values per row, whereas the remaining columns contain the numeric indicators considered significant for evaluation. The estimated range values for the spatial and temporal component of the variogram are listed as *rng_s* *rng_s* and *rng_t*, respectively. Derived from these, the combined quality estimation quotient calculated by Equation 7.7 is given by *rng_qnt*. With *rmse_gn*, also the residuals derived from the Gauss-Newton fitting procedure are considered.

For evaluation of all 108 parameter variants, the RMSE between the reference and interpolated models are plotted against two of the indicators described above as input variables *rmse_gn* and *rng_qnt* in Figure 7.10.

Table 7.2: Result table with systematic evaluation of best 15 out of 108 variogram aggregation variants sorted in ascending order by main quality indicator RMSE.

nr	split_dim	split_pos	aggr_pos	wgt_fnc	rng_s	rng_t	rng_qnt	rmse	rmse_gn
1	max_rel_dev	mea	med	sin	70,61	43,91	1,43	0,35	1,40
2	max_rel_dev	mid	mid	sin	73,10	44,16	1,46	0,35	1,70
3	max_rel_ext	mea	med	sin	68,39	41,41	1,37	0,36	1,58
4	toggle	mea	med	sin	67,62	42,09	1,37	0,36	1,32
5	max_rel_ext	mid	mid	sin	67,24	42,12	1,36	0,36	1,39
6	max_rel_ext	mid	mea	sin	78,42	45,29	1,55	0,37	1,40
7	toggle	mid	mea	sin	80,12	46,25	1,59	0,38	1,43
8	max_rel_ext	mid	med	sin	83,96	46,18	1,64	0,41	1,41
9	max_rel_dev	mid	mea	sin	85,94	48,31	1,69	0,43	1,72
10	toggle	mid	med	sin	86,25	48,17	1,69	0,43	1,45
11	max_rel_dev	mid	med	sin	89,52	49,29	1,75	0,47	1,73
12	max_rel_ext	med	mid	log	96,58	48,02	1,85	0,56	1,22
13	max_rel_ext	med	mea	log	101,25	48,48	1,93	0,63	1,23
14	toggle	med	mid	log	101,72	48,25	1,93	0,64	1,18
15	max_rel_ext	med	med	log	102,46	48,72	1,95	0,66	1,24

As can be seen from both plots, there is no obvious correlation between the RMSE from Gauss-Newton variogram fitting (*rmse_gn*) and the RMSE between the reference and interpolated models (*rmse*). In contrast to that, the quality of estimation of the (joint) *range rng_qnt* strongly correlates with the overall interpolation quality.

In both diagrams we find two clusters where variants have the same RMSE (2.0 and 3.4), which can only be seen in the left diagram where the abscissa values differ. The reason for this is that corresponding methods for aggregation

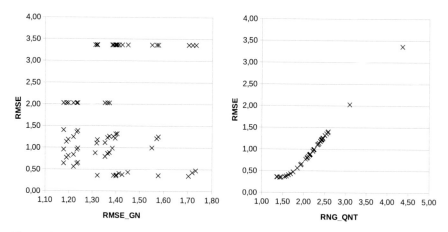

Figure 7.10: Evaluation diagrams of 108 parameter option variants; the RMSE between reference model and interpolation model (ordinate axis) is plotted against the RMSE from variogram fitting by Gauss-Newton (RMSE_GN, left) and the quotient between the compound range derived from variogram fitting and the one from model generation (RNG_QNT, right).

and fitting yield the same upper threshold for parameter estimation, which results in the same RMSE values. Since this upper parameter threshold is inadequate and produces weak results, this effect was not investigated any further.

The most significant property in this experimental series is the weighting function (wgt_fnc). The sine variant appears superior to all other functions, followed by the logarithm-based variant. The first variant that does not use different weights at all (*equ* for equal weights) appears at position 45 out of 108 and thus appears not to be beneficial in any constellation.

For the splitting position (*split_pos*) of the space partitioning algorithm the mean and middle positions perform best, while for the position to be aggregated (*aggr_pos*), the median and the middle position appear beneficial, although not that significantly since the mean variant already appears at the sixth position.

The least distinctive feature in this constellation is the method by which the next splitting dimension (*split_dim*) is determined, since all the three possible variants are among the best four configurations. So the binary space partitioning algorithm appears to group the points into subsets in a way that is not significantly affected by this step according to the properties that are relevant for variogram generation.

Although the experiment in general does not indicate unambiguous advantages for particular variants of aggregation (except for the weighting function), it certainly reveals some preferences that should be considered in forthcoming experiments.

For the most distinctive option *weighting function* there might be potential for further optimization by defining and testing variants similar to the sine or logarithm-based function. This strategy can also be applied to other—actual or future—options, thus evolving towards much better solutions.

7.2.3 Conclusions

The monitoring of continuous phenomena is of high complexity, also because of the interdependencies of the variety of methods and parameters that can be applied [89, p. 42 ff.]. The contribution of the variation module is to facilitate systematic tests and evaluations based on different indicators. The experiment introduced in this section was focused on the crucial task of geostatistics: the variogram estimation.

Although already numerous methodological variations have been evaluated, there is plenty left for further survey resulting from the possible variants of monitoring. If both methodological options and parametric values are varied (e.g. variable x from 3.0 to 8.0 in steps of 1.0), they can be easily included in the variation component.

With the proposed tool, more advanced analyses of the resulting evaluation tables like data mining or steepest ascent [16, p. 188] are possible and might reveal more complex dependencies than the ones identified here.

It has to be stated that the experiment relies on one single synthetic model with a single set of observations. This is, however, a common situation in practice, as Matheron [84, p. 40] points out. However, the results presented here might very well be different for another synthetic model and also for another distribution of samples, so one should be reasonably reluctant from drawing general conclusions from them.

On the other hand, limited generality is the very nature of experimental studies and this does not mean that it is not possible to draw any conclusions at all from them. Rather, they might be considered valid until they are overridden by new experiments that provide deeper insight and more general laws [102], [47, p. 62].

The general architecture of the simulation framework introduced here is supposed to make this process more efficient.

7.3 Sequential Merging

Sequential merging has been introduced in Section 5.4.2 as a technique to exploit the kriging variance in order to conflate several models by weighting their values by their inverse variance [59]. The original motivation for this concept was to mitigate the computational burden of large sets of observations. In (near) real-time monitoring environments where the state model needs to be updated continuously by new observations, the problem of seamless merging can be solved by the same approach.

As a proof of concept, the sequential data merging method is tested experimentally and evaluated according to its performance gain in this section.

7.3.1 Experimental Setup

In order to prove the feasibility of the approach, but also to reveal its impact on accuracy, a simulation with appropriate indicators is performed. For this purpose, a synthetic continuous field is used. It is derived by kriging over 14 rain gauge stations and depicted in Figure 7.11(a).

At this stage, we ignore temporal dynamism in order to exclude it as a factor for differences (RMSE) between the reference model and the sequential approach. In the simulation scenario, the continuous grid model serves as reference. Random observations are scattered over the model area, each assigned the value picked from the reference model at its position. Given this simulated measurement set, a new model can be calculated by kriging (Figure 7.11(b)).

The derived model (Figure 7.11(b)) differs from the reference model (Figure 7.11(a)) due to interpolation uncertainty, but approximates it well when the number and distribution of samples are sufficient (see Section 5.3.2).

Following the sequential strategy, subsets of the synthetic measurements are created and calculated sequentially in sub-models (see Figures 5.15 and 5.16 in Section 5.4.2).

For the first subset (Figure 7.12(c)), the deviations to the reference model (a) are rather large and can be seen in the difference map (d). Calculating all subsets of the data and merging them successively by weight leads to the final model (e), which also considers all the sample data, but unlike model (b) in a sequential manner.

The difference map (f) expresses the discrepancy towards the reference caused by the sequential approach. The overall discrepancy per model can be quantified by the root-mean-square error (RMSE) relative to the reference model (a). In the following, this value is used to indicate the fidelity of these interim models.

7.3.2 Results

In Figure 7.12, the computing time is plotted against the RMSE relative to the reference model for both the complete model calculation (square) and the sequential method (connected dots). Randomized sets of points (100, 200, 300 and 400) were subdivided into subsets or sub-models of 10 points each.

As can be seen from the results, the sequential method has a lower accuracy in total, but provides a coarse result almost immediately. Within each plotted scenario, the RMSE tends to decrease when following the sequence. The $\mathcal{O}(n^3)$-effect of the conventional calculation becomes obvious when comparing its total computing time to the one of the sequential approach for a large n.

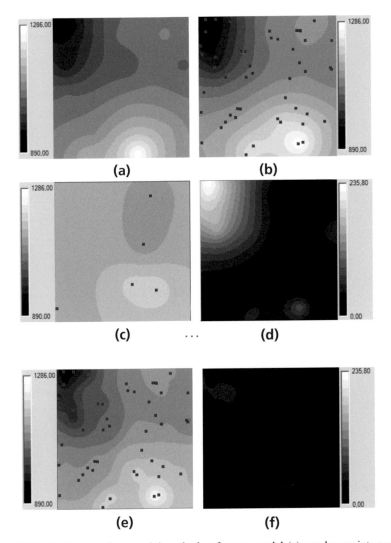

Figure 7.11: Evaluation of sequential method: reference model (a); random points and corresponding model derived from those points (b); first subset of random points (c) with corresponding difference map (d) towards reference model (a); sequentially updated model of all subsets (e) with resulting difference map (f) towards the reference (a).

Since the observations are distributed randomly over the reference model, the results also tend to scatter when the scenario calculation is repeated. But the general behaviour of the algorithm is reproducible in essence.

The tests introduced here are designed to explore the general behaviour of the approach. It converges to a saturation value and for large models clearly outperforms the conventional method in computing time.

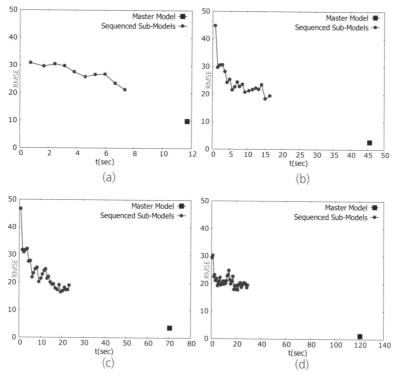

Figure 7.12: Performance comparison between master model (square) and sequenced calculation (dots): (a) 100 samples, (b) 200 samples, (c) 300 samples (d) 400 samples; subdivision is done in a way that sub-models contain 10 samples each.

7.3.3 Conclusions

This experiment was designed to demonstrate the general feasibility of a sequential strategy when performing kriging interpolation. It exploits the kriging variance as a continuous weighting schema of the models to be merged. The approach addresses the computational complexity of kriging for large datasets and the problem of integrating new observations into an existing model, as present in real-time monitoring scenarios.

The results show that the approach reduces total computing time for large datasets and provides coarse models immediately. It defines a rule for seamless merging of partial models based on the information confidence level given by the variance map. For real-time monitoring systems that are fed by a continuous data stream, the method provides fast responsiveness and can adapt to data load and available resources.

For real monitoring scenarios, this method will have to be refined to generate acceptable results under given circumstances like data stream characteristics, model update intervals, computational resources and quality requirements. With the framework proposed, such circumstances can be considered in simulation scenarios. Different method variants and parameters can be applied and evaluated by using appropriate output indicators. Feasible approaches and settings can thus be identified before using them for real monitoring scenarios.

7.4 Compression

In this section, the compression algorithm as described in Section 5.4.3 is applied to empirical data. After an introduction of the specific data format of drifting buoy data, an evaluation of the achieved compression rate is carried out.

7.4.1 Experimental Setup

For the evaluation of the compression algorithm, data from the *Argo* drifting buoys program were used[1]. The format was provided by a Canadian governmental service[2]. It is used for the experiments and is described in Listing 7.1.

```
Contents:
Col 1 = Platform identifier (ARGOS #)
Col 2 = EXP$ - The originator's experiment number
Col 3 = WMO$ - WMO platform identifier number
Col 4 = Position year/month/day hour:minute (UTC)
Col 5 = Latitude of observation (+ve North)
Col 6 = Longitude of observation
        (+/- 180 deg +ve West of Greenwich)
Col 7 = Observation year/month/day hour:minute (UTC)
Col 8 = SSTP - Sea surface temperature (deg. C)
Col 9 = Drogue on/off - 1 = attached; 0 = not

Note: Missing value indicated by 999.9999
```

Listing 7.1: Original header of ARGO drifting buoy data.

The sample contains all data types mentioned in Section 5.4.3. We find *Integer* types for the IDs in columns 1, 2 and 3. Colums 4 and 7 contain *DateTime* types, columns 5, 6 and 8 represent *Double* numbers while column 9 displays an on/off state as *Binary*.

From the original dataset, subsets of 100, 1000 and 10000 points are selected by spatial and temporal bounds. Listing 7.2 depicts a corresponding data header generated by the compression algorithm (note the changed names and order compared to Listing 7.1). The values for *min* and *max* are derived from the actual data. Together with the preset value *max_dev* for the maximum deviation, the bit depth is determined using Equation 5.20. The value *bpr* indicates the number of *bits per row* used for each column.

---

[1]http://www.argo.ucsd.edu, visited 2018-02-19.
[2]http://www.meds-sdmm.dfo-mpo.gc.ca/isdm-gdsi/drib-bder/svp-vcs/index-eng.asp, visited 2016-04-27.

A maximum deviation of 0.5 for integers means that at full bit depth the exact number is provided. For the *DateTime*-type this value represents seconds, so the minutes are decoded accurately when it is set to 30.

```
fname  max_dev bits bpr  min                     max
x       0,0005  16   1    10,767                  49,671
y       0,0005  15   1    40,07                   59,08
val     0,0005  14   1    5,529                   18,55
idarg   0,5     16   1    37411                   92885
idexp   0,5     12   1    6129                    9435
idwmo   0,5     23   1    1300518                 6200926
tpos    30      10   1    2010-12-31 21:54 2011-01-01 09:24
tobs    30      10   1    2011-01-01 00:05 2011-01-01 09:57
drg     0       1    1    False                   True
```

Listing 7.2: Header for the compressed dataset of ARGOS drifting buoy observations.

As can be seen from Listing 7.2, the value for *idwmo* has the highest bit depth of 23, since the range of that value is nearly five million. The effect is that the longest chain of bits occurs for that dimension in the corresponding data file (see Listing 7.3).

```
     iii             iii             iii
    dddtt           dddtt           dddtt
   vaewpod         vaewpod         vaewpod
   arxmobr         arxmobr         arxmobr
  xylgpossg       xylgpossg       xylgpossg
  ---------       ---------       ---------
  101011001       010101000       111101000
  10100000        01110000        00010000
  01001100        00100100        00000100
  11000010        00010010        00010010
  01001000        01001000        00101000
  01100000        01111000        01011001
  01101000        01001001        00101000
  10000100        01101100        11101101
  01000110        11000111        10000111
  11100100        00010100        11110100
  001001          101011          111011
  000001          010111          011111
  0100 0          0101 0          1010 0
  1010 1          0111 1          0001 1
  01 1 1          10 0 1          10 1 1
  1  0  0         1   0 1         0  1 0
       1               0               1
       1               0               0
       1               1               0
       1               0               0
       1               0               0
       1               0               0
       0               1               1
```

Listing 7.3: Compressed data for three observations of ARGO drifting buoys (column names and values to be read vertically).

Three observations are listed, each containing all nine data columns organized vertically (as are the column names) with increasing accuracy from top to bottom. As can be seen, the binary value of the rightmost field *drg* (indicating drogue on/off) is already complete in the first row whereas the one for *idwmo* is resolved in row 23, as indicated in the header file (Listing 7.2).

Since this column represents an ID, it might be necessary to resolve it earlier than in the last data row in all likelihood. Therefore, the number of bits per row is increased to four. The resulting structure for the same data can be seen in Listing 7.4.

```
iii              iii              iii
ddd    tt        ddd    tt        ddd    tt
vaew   pod       vaew   pod       vaew   pod
arxm   obr       arxm   obr       arxm   obr
xylgpo ssg       xylgpo ssg       xylgpo ssg
------------     ------------     ------------
100011010001     010101010000     101011010000
10000000100      01000000100      11100000100
01001111100      00010111100      00101111100
01000011000      00010011100      00100011000
11001111100      01011001000      01101010110
11000110 00      01011001 00      11100100 00
11101    10      01011    10      11101    10
10000    10      11111    10      01100    00
11110    00      11010    00      10100    11
11000    10      10100    11      01100    11
00010    10      10011    11      01100    11
10110    00      01101    01      00100    01
1111             1110             0010
0010             1101             1110
0110             0001             1110
1 10             1 11             1 10
 0                0                1
 1                1                1
 1                1                1
 0                1                1
```

Listing 7.4: Compressed data with prolonged bit length of four per row for column *idwmo* (column names and values to be read vertically; data columns without name belong to *idwmo*).

In this configuration, the exact values for *idwmo* are already resolved in the sixth row since the three columns to the right without title are also utilized. In practice, the bit length per row can either be set directly (column *bpr* in the header), determined by maximum number of rows, or by some arbitrary combination of accuracy and row number in the form "accuracy *x* must be met in row *n*". This configuration can be set individually for each dimension to achieve a good balance between stepwise accuracy improvement and total size per data row.

7.4.2 *Results*

To create indicators for the performance of the compression method, it is applied to a dataset of 100, 1,000 and 10,000 observations given in the format described above. We compare four indicators here: The first indicator is the size of the text file as received from the Canadian governmental service provider (denoted by "Text" in the following).

The necessary space when the data is parsed and translated into native machine data types is evaluated as a second indicator ("Native"). We assume 32 bits for *Integer*, 64 bits for *Double*, 64 bits for *DateTime* and 8 bits for *Boolean*. The proposed binary format of the BSP compression algorithm is the third format listed. The size of the header is not considered here. Finally, a ZIP compression of the text file is applied as the fourth format with 7-Zip[3] using following settings: normal compression level, deflate method, 32 KB dictionary size and a word size of 32.

As can be seen in Figure 7.13, the approach outperforms the ZIP compression for the small sample. With growing data size, the efficiency of the ZIP dictionary is increasing, which is not the case for our approach. Nevertheless, taking into account progressive decoding as a key feature, the slightly worse compression ratio for large datasets appears acceptable.

Remarks on Reasonable Extensions

The proposed compression method as introduced so far fulfils the requirements mentioned at the beginning of this section. There are, however, some ideas not yet implemented but certainly worth considering to be realized in future.

In the sample buoy data introduced in Section 7.4 we find missing measurement values indicated by "999.9999" (see header in Listing 7.1). The idea behind this number is to have an optical pattern immediately recognizable for the human eye as exception. Using it as "unset"-indicator within the compression algorithm is rather awkward, since, by being an absolute outlier, it enlarges the value domain (and therefore the necessary bit depth) significantly. A more explicit variant is desirable here, e.g. by indicating validity/invalidity of a value by its first bit. In case of invalidity, the bits to follow for that particular dimension can simply be dropped, but on the other hand, this mechanism would only pay off when having a significant amount of unset values.

Another aspect worth considering is the compression of the header associated with each compressed data segment (see Listing 7.2). With its metadata for each value dimension (name, deviation, bit depth, bits per row, min, max) it is crucial for archiving and retrieval and a prerequisite for correct decoding. In a monitoring scenario with very small data segments to be compressed, the relative size of that header can justify its compression where transmission of data is expensive.

---

[3]`https://www.7-zip.org/`, visited 2020-06-18.

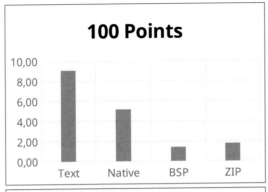

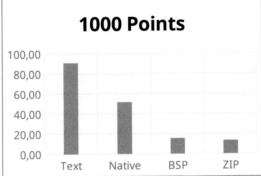

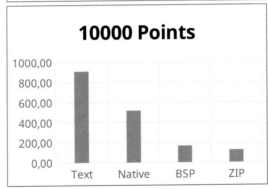

Figure 7.13: Data volumes (KB) in different formats for 3 datasets.

Wherever the transmission of the compressed data is potentially error prone and not secured by other protocols, it might become necessary to implement some checksum method within the binary format itself. In this case, the gain in reliability needs to be carefully weighted against the expenses according to implementation, processing and data volume.

7.4.3 Conclusions

The methodology presented here is useful for situations where massive sensor data needs to be compressed in a way that allows a progressive retrieval with increasing accuracy per step. It supports the most typical data types found in sensor data like *Float*, *Double*, *Integer*, *Boolean*, and *DateTime*, each one with specific compression schemata. The compression ratio depends on the value range and necessary accuracy. The number of bits per transmission step can be set in accordance with the transmission priorities, e.g. if particular dimensions are needed with higher convergence of accuracy per step.

In a wireless network scenario, the method would require some overhead for communication between data nodes. For example, the header that determines the mode of transmission needs to be exchanged before transferring the data. In environments where transmission of data is significantly more expensive than processing coding and decoding tasks, the method is likely to pay off.

For using the proposed method in a real-time environment, some protocol needs to be created to retain efficiency of transmission: values of defect sensors can be omitted, changed value ranges need to be adjusted and maybe the bits-per-row configuration shall be changed due to changed priorities. All this means considerable overhead which must be weighted carefully against achievable savings for data transmission.

When thinking about long-term archiving of data streams in databases, there are several points to be considered. Maybe the most important one is how a large dataset is to be segmented into smaller units. Doing this by spatial, temporal or spatio-temporal boundaries is reasonable since this is the most obvious means to refer the sensor data to other aspects like e.g. traffic density. Databases today widely support efficient management of spatial and spatio-temporal data [17].

But the associated indexing techniques were primarily developed having retrieval performance and not compression in mind. Thus, it appears reasonable to make use of them at a higher granularity level than the individual observation. So the method proposed here can be applied to appropriate segments of data while using the spatial or spatio-temporal boundaries of these segments for indexing with common database techniques. The compressed segment can be stored as large binary objects (BLOBs) in the database with the associated spatial/spatio-temporal index and metadata.

Since the spatio-temporal boundaries can also be seen as statistical properties of the dataset, it is reasonable to ask if additional statistical properties like mean value, standard deviation or skewness should not also be considered for each dataset. This might be of little use for the dimensions space and time, but can be crucial for measured values like temperature or air pollutants. If advanced analysis methods like geostatistics are used, more complex statistical indicators like variogram model parameters should be integrated [80]. All this data should be stored as metadata alongside each dataset to support efficient retrieval.

One central issue here is the way how large datasets are subdivided into smaller subsets on which the compression method is applied to and the corresponding metadata are related to. A good configuration balances retrieval granularity, subset management overhead, indexing costs, transmission data volume, system responsiveness and accuracy in a way that fulfils the requirements of the whole monitoring system.

7.5 Prediction of Computational Effort

The general idea behind the experiment set out in this section is to compare the computational effort predicted by the model approach of machine-independent description from Section 5.5.2 with the one actually measured in experiments. Different hardware with different configurations according to multithreading are applied to test the generality of the concept.

7.5.1 Experimental Setup

To evaluate the concept from Section 5.5.2, it was applied to the resource intensive process for generating continuous random fields. The computational workload of this algorithm, obtained with help of the function *QueryProcessCycle-Time* and expressed by the metric *gigacycles*, increases when increasing the *range* value of the associated variogram, since more grid cells have to be considered to calculate the weighted mean.

Experiments were carried out for different modes on two different CPUs: an Intel® Core™ i3-5010U CPU with 2.1 GHz and 4 logical processors (2 cores) and an Intel® Xeon™ E5-2690 v3 with 2.6 GHz and 24 logical processors (12 cores).

In the study, the multithreading overhead factor *thro* from Equation 5.21 was quantified to 0.5 for both processors, which means that, in this case, the gain in performance in effect coincides with the number of *physical* (not *logical*) processors of the used CPU. This indicates that the multithreading functionality within one core can rarely be exploited for this task. For the Xeon™ CPU, the frequency correction factor fc is set to 1.5 to express its apparently better instructions-cycle ratio.

By switching the parallelization mode on and off for the critical loop in the algorithm, we get four configurations to evaluate according to the proposed model for estimation of computational effort. The cycles counted by the Intel™ Core™ i3-5010U CPU in the singlethreaded mode are used as a reference for Equation 5.21.

In the given process, there is one portion of code that can only be processed as a single thread because it contains routines difficult to parallelize. The other portion contains the critical loop executing the moving average filter by numerous iterations. Because of its high workload impact, this loop has deliberately been optimised with respect to parallelization.

7.5.2 *Results*

Based on the proposed metric, Figure 7.14 refers the time expense predicted by the equation (lines) to the time actually needed for calculation (points). The *calculated* time effort represented by the lines is composed of the sum of algorithmic portions for each workload position on the abscissa.

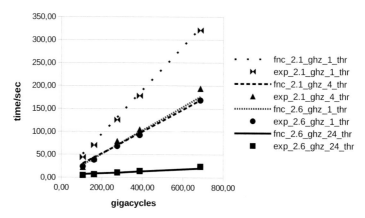

Figure 7.14: Performance evaluation of four computer system configurations: (1) 2.1 GHz with 1 thread, (2) 2.1 GHz with 4 threads, (3) 2.6 GHz with 1 thread and (4) 2.6 GHz with 24 threads. The lines (prefix *fnc*) represent the prediction by *function* for each configuration and the points (prefix *exp*) represent the *experimental* data.

The plot clearly reveals the scaling effects of multithreaded processing, which, as already stated in Section 5.5.2, is the crucial leverage for contemporary performance improvement. It can also be seen that predicted and actual time expenses have similar values. Only the variant with 2.1 GHz and 4 threads does not scale as well as predicted. One or some of the blurring influences mentioned in Section 5.5.2 can be assumed to cause this effect.

7.5.3 *Conclusions*

The experiment was focused on the generic quantification of computational workload in order to estimate the temporal effort that is necessary on different platforms. The evaluated toolset is capable of estimating the processing time of complex calculations for different configurations of scenarios of monitoring, analysis and simulation.

In combination with the toolset for systematic variation and evaluation (see Section 5.5), this approach allows for deep analysis of multiple constellations according to methods, parameters and hardware configurations and their effects on performance indicators. Feasibility and efficiency studies for different configurations can thus be carried out without actually using the intended hardware. A calibration of the parameters of Equations 5.21 and 5.22 might be necessary if they cannot be sufficiently obtained from hardware specifications.

The central concern of the concept is to abstract from concrete hardware configurations, algorithmic parameterizations and output indicators and thus to simplify and standardize comprehensive evaluation scenarios. Strategies for incremental optimization are thus made explicit and transparent. This can significantly help to make experimental results well documented and reproducible.

The modelling and consideration of data transmission costs, which is crucial for wireless networks, was only briefly mentioned as one possible factor of optimization. An inspiring scenario in this context would be a complex simulation with distributed sensors and full knowledge about geometric constellation and profiles for data transmission expenses that depend on that constellation. Placing the sensors on autonomous platforms increases complexity and imposes extensive simulation to find strategies for a high overall efficiency. The tools presented here could be of significant importance to master the complexity of such missions.

7.6 Higher Level Queries

As has been outlined in Chapter 6, a fundamental feature when monitoring continuous phenomena is to define a condition that has to be checked against steadily by an ongoing process of observation, interpolation and aggregation. In many cases, it might be reasonable to continuously falsify a hypotheses like an exceeded threshold for a particular air pollutant. Thus, the monitoring system will trigger a notification in both cases: (1) when the threshold is actually exceeded and (2) when the observations actually made do not suffice to exclude the possibility of this dangerous situation.

Given the experimental framework as used for this work, there is a corresponding test setting that can be established in order to represent the scenarios as described above. It will be specified and assessed in the following.

7.6.1 Experimental Setup

The general experimental infrastructure that is extensively relied upon by this work can be summarized by the following features (see also Figure 5.3 on page 69):

- provision of a continuous random field according to a particular variogram model
- generating randomly placed observations as (multi-dimensional) point set within this reference field
- creating the experimental variogram by relating the halved squared differences (semi-variances) of two observations to their distances in space and/or time
- aggregation of the experimental variogram for a more efficient parameter estimation
- estimation of the variogram parameters by fitting the function to the aggregated points of the experimental variogram
- interpolation of a grid by kriging
- error assessment by deviation maps and the total RMSE against the synthetic reference model

This process chain was primarily designed in order to evaluate the monitoring process. The model error is expressed by the RMSE as a single and expressive quality indicator. By systematic variation of algorithm parameters within this process chain it is possible to yield that one configuration which performs the best in "reconstructing" the reference random field. It is basically the method pursued in the context of this work in order to evaluate the introduced approaches.

In this section, however, the focus is not on the overall interpolation quality, but rather on the formal expression and assessment of complex states in the context of continuous phenomena. As already outlined in Chapter 6, the idea for a basic feature of a monitoring system is to continuously reject the hypothesis that there is a hazardous environmental state.

For such a hypothesis rejection, the analysis can not be limited to the interpolated and aggregated values of the region of interest. This would potentially approve a situation as safe where there are only a few observations that by accident are placed at safer spots within the otherwise more hazardous region of interest.

Interpolation between those spots will eventually lead to a false representation of the region. Consequently, any aggregation generated from this will also provide false non-critical values.

To avoid such situations, the sufficiency of the observations has to be an integral component of the hypothesis test to be performed.

As an outstanding feature among interpolation methods, the kriging variance (see Section 3.5) can very well serve as a measure to assess this sufficiency. While providing an estimation of variance for each interpolated value, it can represent an areal quality indicator when it is aggregated.

To demonstrate the expressiveness of the kriging variance, two experimental process chains as described and depicted in Section 5.3 were executed. The focus of the evaluation is on the assessment of a particular region of interest.

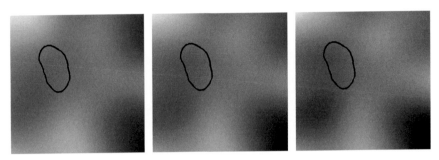

Figure 7.15: Aggregation region within a continuous random field; the marked region is to be aggregated to estimate the average value of the observed phenomenon over time.

Figure 7.15 shows 3 of 10 frames of a model that has been generated by a variogram filter (in this case with a spatial range of 50 m) as described in Section 5.3.1. The general parameters of the random field that is used as reference model for two experimental series are:

- grid size of 150 x 150 (spatial) x 10 (temporal) elements (=225,000 grid cells)
- spatio-temporal extent: 150 m x 150 m x 60 min
- white noise field with mean of 5000 and deviation of 500
- interpolation: kriging by separable covariance function, parameters fitted by Gauss-Newton
- aggregation region: 13,620 grid cells (approx. 30 m x 50 m on 10 frames)

In order to test the aggregation for more than one phenomenon dynamism, two different experimental series were carried out. Primarily, they differ in the spatial range (30 m and 50 m) of the variogram filter used to generate the random field. Consequently, the model based on the variogram model with shorter spatial range is more dynamic and therefore has to be observed with higher density (see Subsection 5.3.2). So for the two experimental series, the parameters that differ are:

■ filter: separable covariance function based on spherical function for spatial and temporal dimension, spatial range of 50 grid cells ($\triangleq$ 50 m for series 1) and 30 grid cells ($\triangleq$ 30 m for series 2), temporal range of 30 grid cells ($\triangleq$ 30 min) for both series

■ sampling for series 1: 10 to 200 samples in steps of 10 (the minimum number according to Equation 5.8 is 144 samples)

■ sampling for series 2: 50 to 500 samples in steps of 25 (the minimum number according to Equation 5.8 is 400 samples)

The main goal of the experiment is to evaluate the representativeness of a monitoring setting, as outlined in Chapter 6. Therefore some output indicators that are considered meaningful in this context are inspected. Following indicators are derived from the synthetic model and serve as a reference for the indicators as derived from the individual interpolated models:

■ mean of grid cell values within the region of interest
■ variance/deviation of grid cell values

The following indicators are derived from each monitoring series item, ranging from 10 to 200 samples for series 1 with spatial range of 50 m, and from 50 to 500 samples for series 2 with spatial range of 30 m, respectively:

■ number of samples used for interpolation of series item (abscissa)
■ mean value of interpolated grid cells within the region of interest
■ value deviation as an indicator of the dynamism within the region of interest
■ root mean kriging deviation
■ RMSE of the entire interpolation model against the entire synthetic reference model

The root mean kriging deviation expresses the overall model variance. It is aggregated from the kriging variance that is provided for each interpolated grid cell. Since the mean value is *not* calculated as a *weighted* mean (although in principle feasible given the kriging variance for each cell), its variance can not be determined by Equation 5.18 (p. 95). Instead, it is the root of the summed squared deviations (here: kriging variances) in analogy to the RMSE.

A *weighted* mean value aggregation does not make sense here since it would bias the estimation towards denser observed regions. But the interpolated values do not represent independent and identically distributed (IID) observations of *one* mean value, but are rather estimations of the mean of multiple discrete values within a continuous field.

Instead of a classical statistical variable we deal with the more complex subject of a gaussian process [106] or second-order stationary process [31] here. This needs to be considered when interpreting the results of the experiments.

7.6.2   *Results*

In Figure 7.16 and Figure 7.17, the parameters listed in the last subsection are related to each other in a series of experiments with an increasing number of samples. The series of experiments was carried out in a way such that each sample set contains the samples from the preceding one and adds new observations to it. So within one series, the interpolation generated from 70 samples contains all samples from the one with 60 samples and adds 10 new ones and so forth. Otherwise, the effect of improved estimation by increased amount of observations might be partially blurred since observations are placed at random in this experimental setting.

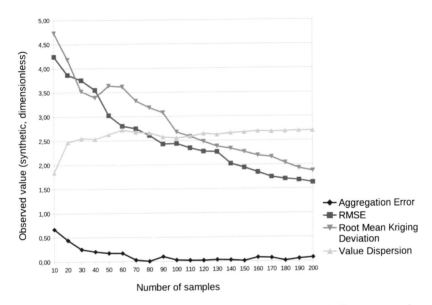

Figure 7.16: Aggregation evaluation with spatial range of 50 m. Four indicators are evaluated for each series element.

As expected, the *RMSE* decreases with denser sampling as it improves the overall model accuracy. A similar trend can be seen for the estimated overall model accuracy derived from the kriging variance, the *root-mean kriging deviation*.

For the crucial value of interest here, the error of the estimation of the mean value (named *aggregation error*), a saturation can already be found at 70 samples (Figure 7.16) and 225 samples, respectively (Figure 7.17).

The *value dispersion* is determined analogously to a standard deviation, but does *not* represent the flaws caused by insufficient observations or interpolation; it is instead an indicator of the dynamism of the phenomenon itself. As can be

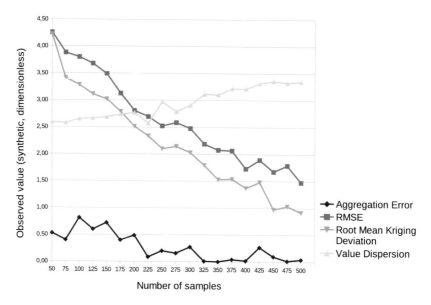

Figure 7.17: Aggregation evaluation with spatial range of 30 m. Four indicators are evaluated for each series element.

seen in the diagrams, this dynamism is underestimated for small sample sizes and converges towards the value derived from the reference model, which is 2.82 for the first experiment and 3.50 for the second.

This effect is comprehensible since sparse sampling will generate surfaces that are smoother than the actual phenomenon. On an average, the model might still provide the best estimate for each position, but eventually it does not reflect the overall dynamism very well. When the density of observations increases, the dynamism of the phenomenon is captured more accurately. This effect is more striking in the series with the shorter spatial range of 30 m (Figure 7.17), since due to the more dynamic surface there is more potential for improvement than for the rather plane surface.

The need for the number of necessary samples for the given model, according to Equation 5.8, is met by 144 and 400, respectively. The experiments show that approximately from these points on the abscissa there is no significant improvement in terms of the estimation of the dynamism of the phenomenon as expressed by the *value dispersion*. For the error with estimating the mean value itself (*aggregation error*), this is already the case with significantly fewer samples.

So the data indicates that for capturing the dynamism of a phenomenon, more samples are necessary than for estimating the mean value, which seems reasonable in this context.

The dynamism of the phenomenon as expressed here by the deviation of values is also associated with the theoretical variogram. However, while the latter entails more complex information, the former is a rather coarse indicator for dynamism. It was also used here to represent a small region within the observed area which might not be adequately represented by the variogram that represents the whole area. This circumstance might therefore justify its utilization in scenarios similar to the one described here.

7.6.3 *Conclusions*

The most important insight from this study is the apparently strong correlation between the estimated deviation (expressed by the root-mean kriging variance) and the actual deviation expressed by the root-mean-square error (RMSE) between the reference model and the interpolated model.

This assessment has important implications for situations where conclusions have to be drawn from potentially insufficient observations: it is a metric by which the reliability of aggregation-based statements like the mean value about a continuous phenomenon can be estimated. Notwithstanding the very limited experimental coverage associated with this problem as presented here, the results appear convincing enough to consider the proposed method as an adequate approach to tackle such uncertainty. It can be helpful to decide whether or not the assertion "The mean exposure with particulate matter in district X for June is below $50\,\mu g\,m^{-3}$ PM_{10}" can be stated confidently.

It is important to emphasize that the estimation of the kriging variance depends on the quality of the variogram estimation (see Section 5.3.3, 5.3.4, 5.3.5). So either the sampling needs to be dense enough to generate a representative variogram, or the variogram is known prior to the processing of the sensor data. This might be the case for routinely performed monitoring of phenomena that are well understood.

Once the variogram is determined sufficiently well, the kriging variance depends only on the distribution of samples [54, 45] and so does the *aggregated* kriging variance. So areas which are rather poorly observed will indicate this by a high aggregated kriging variance, which might be a crucial hint for decision making. The process chain for this complex aggregation is outlined in Figure 7.18.

As the schema in Figure 7.18 illustrates, the aggregation of interpolated grid values involves an important concept to be considered carefully: the combined utilization of the value grid and the variance grid, or "map of the second kind" [89], in order to increase the value of the information in this condensed form.

Table 7.3 lists the meaning of the two aggregations (mean and variance/deviation) when applied to the value grid and to the kriging variance grid, respectively. The properties of the table entries will be explained in detail in the following while referred to by their numbers.

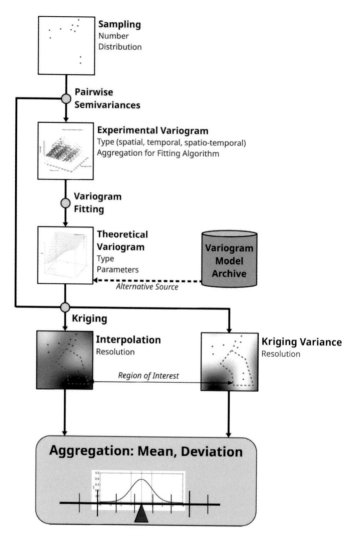

Figure 7.18: Aggregation process chain: The experimental variogram is generated by combining samples pairwise and plotting the squared halved differences in value (semi-variance) against the spatial and/or temporal distances. The theoretical variogram of a specific type is fitted to those points, eventually after they have been aggregated in order to unburden the optimization algorithm. The theoretical variogram might also be obtained from an archive comprising previous monitoring or other scientific data. Based on this variogram model and the actual observations, the kriging algorithm generates the interpolation grid and the variance grid with the intended resolution. For the region of interest, the aggregation is generated from these grids. Besides the value itself, it also provides a deviation indicator derived from the aggregated kriging variance.

The grid related to the interpolation values should represent the best information about the phenomenon that can be derived from the available observations. If

Table 7.3: The aggregation indicators (*mean* and *variance/deviation*) of the two grids (*interpolation value* and *kriging variance*) within the region of interest represent basic properties of the phenomenon.

	Mean*	Variance/Deviation
Interpolation Value	(1) Value of interest	(2) Observed phenomenon dynamism
Kriging Var./Dev.	(3) Overall model accuracy	(4) Homogeneity of sampling

* In the case of kriging variance, the mean is calculated as the root of the summed variances, in analogy to the RMSE (see p. 157).

properly managed, it provides this information instantly for each position without any further interpolation efforts. Thus, it can be used for visualisation or intersection with other spatial or spatio-temporal data. A basic application in this context is the intersection of the grid with an area of interest for which the involved grid cells shall be aggregated.

(1) The mean value expresses the aggregated property of a confined area with respect to the observed variable by a single numeric indicator. A use case for this approach might be the estimation of the total exposure of the human body while residing in a hazardous area.

(2) On the other hand, the variance or deviation of the interpolation value expresses the dynamism of the phenomenon as it is represented by its model. For example, this value might indicate if there is significant variation of exposure within the region of interest. This could be helpful to determine whether the assignation to this region alone is a sufficient exposure indicator. Consequently, it might call for a finer modelling granularity in space and/or time.

Besides its undisputable interpolation quality, the method of kriging generates the *kriging variance/deviation* as valuable additional information. By estimating the interpolation error at each position, it allows for versatile application options.

(3) The mean value of the kriging variance (or *root mean kriging deviation*) within the observed area expresses the general accuracy of the model. The experiments show a strong correlation between the aggregated kriging deviation and the deviation of the interpolated model from the reference model as given by the RMSE.

(4) The variance of the values in the kriging variance grid, or, so to say, the variance of variances, can be seen as representing the homogeneity of observations. It might indicate that sensors within a region should be distributed differently in order to utilize the given resources more efficiently.

As the experiments show, there are strong arguments in favour of the utilisation of kriging and the kriging variance that comes along with it. Beside the sophisticated, differentiated and highly adaptive way of interpolating the value from observations, it provides a very valuable indicator in monitoring scenarios: the confidence of that interpolation.

Section 5.4 already introduced two applications of the kriging variance: (1) a "smart" filtering of continuous observations before interpolation and (2) a sequential model merging method to integrate new observations into an already existing model.

The rationale behind the approach described here is to support one substantial goal of monitoring per se: the detection of hazardous environmental states. In order to ensure this feature reliably, a system needs to be designed in a way that also considers the danger of insufficient observation or technical failure.

The guiding principle that was applied here is the reversal of the burden of proof: instead of constantly checking for an exceeded threshold, the method sketched here supports the paradigm of constantly proving the absence of such a situation. This deliberately covers cases where failed sensors might lead to hazardous states remaining undetected.

A monitoring devoted to this paradigm will constitute an intuitively manageable and therefore much more reliable system. If implemented consequently, it incorporates a mechanism to uncover weaknesses in the arrangement of observations and thus constantly provides reassurance of its functionality.

7.7 Case Study: Satellite Temperature Data

The experiments introduced so far were carried out on synthetic models generated by a filter (see Section 5.3.1) or on models derived from interpolated observations (as in Section 5.4.2). The idea behind this approach is to have full knowledge about the observed phenomenon.

Besides this, working with synthetically generated models has the advantage that the variogram parameters estimated from the observations (see Section 5.3.5) can be compared to the ones known previously. Interdependencies between the quality of parameter estimation and overall interpolation quality can thus be identified (see Section 7.2). Furthermore, the formula for the minimum sampling density (Equation 5.8) can thus be tested for different constellations of dynamism.

To also apply the framework to *empirical* data, a remote sensing thermal image, obtained from the National Oceanic and Atmospheric Administration (NOAA), is used as a reference in this section. In practice, such imagery data usually does not have to be interpolated since it already contains the variable of interest in the required resolution (except for occlusions, e.g. by clouds [117, 31]). So the advantage of knowing the complete model in the given resolution is also given here. What is *not* given—in contrast to the synthetically produced model—is any prior information about the dynamism of the observed phenomenon.

7.7.1 Experimental Setup

As with the synthetic data, in this experiment the given raster image is also sampled with random observations which are used to estimate the unobserved grid cells. Analogously, the accuracy of the interpolation can then be quantified by the difference between the reference and the derived model.

As a reference model, a satellite raster image of the *4 km Pathfinder SST Climatology* provided by the National Centers for Environmental Information (NCEI)[4] from the National Oceanic and Atmospheric Administration (NOAA), was selected. It provides the sea surface temperature with a ground resolution of about 4.6 kilometers (for both latitude and longitude, since a region in the Atlantic Ocean near the equator at $4°S, 12°W$ was chosen). The image entails 150 rows and 150 columns, resulting in 22,500 grid cells. An area of about 697 * 697 km is covered. The image was taken at night on 2013-01-01. The temperature is stored as a 32 bit floating point value of unit kelvin.

Figure 7.19 displays the grid using grey scaled values.

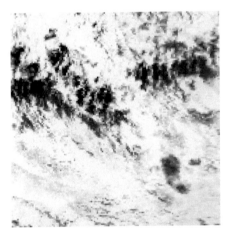

Figure 7.19: Sea surface temperature (SST) satellite image extracted from the 4 km Pathfinder SST Climatology provided by the National Centers for Environmental Information (NCEI). The values range from 284.0 (bright) to 298.9 (dark) K (10.8 to 25.7°C, respectively).

Assuming a sufficient signal-to-noise ratio, the grid represents the sea surface temperature (SST) at the given resolution. Although a predominantly continuous character can be granted, there are discontinuous patterns as well, especially in the upper half of the image. While the overall standard deviation of the tempera-

---

[4]https://www.ncdc.noaa.gov/cdr/oceanic/sea-surface-temperature-pathfinder, visited 2018-02-19

ture grid is 3.3 K, we find differences of more than 10 K for neighbouring cells at the edges of these patterns.

These effects are caused by ocean circulation and oceanic fronts [123, 110]. Sun et al. [117] extensively cover this issue by explicitly incorporating such fronts in their interpolation model. The method is proposed for applications where discontinuities are rather common, like for oceanography or soil moisture monitoring. Gaps in observational coverage, e.g. caused by clouds, are thus bridged by patterns that entail such fronts.

The subject of discontinuities is beyond the scope of this work. Instead, the set of methodological and parametrical variations given by the framework are applied to the image in order to identify the best configuration from the given set of variants for this particular kind of phenomenon.

The satellite image indicates the superiority of the remote sensing method. In this example, it provides a gapless and consistent representation of the phenomenon. Nevertheless, remote sensing is not always available due to clouding, coverage, cost, among other factors [110]. Moreover, not every phenomenon can be captured sufficiently by imaging techniques like remote sensing (see also Section 1.3).

Consequently, discrete sensor observations, although maybe very sparse, are often the only source of knowledge that can actually be obtained. Even if the estimation of a value by kriging interpolation between those sparse observations might not be precise at all, it often provides the best results—according to bias and variance—that can be generated from these observations [29, p. 239], [95, p. 88].

Notwithstanding the presence of partial discontinuities, the necessary sampling density for the image can be estimated by Equation 5.8. Since it entails the *range* parameter, which is not known beforehand for this dataset, it was estimated iteratively to be about 22 grid cells. With Equation 5.8, we get 186 observations as minimum density, which was rounded up to 200 randomly dispersed observations for the experiment. This results in 19,900 variogram points (see Section 3.2) by pairwise combination.

It has to be stated, though, that the equation was deduced for phenomena that are considered stationary, which is not strictly the case for the given dataset. Depending on the aim of the interpolation, this approach might nevertheless very well be reasonable, as will be discussed with respect to the results below.

As already mentioned, the idea of the experiments described in this section is to treat the grid as derived from satellite observation as a reference, just like the synthetic random fields are treated in Section 5.3.1. Analogously, a random set of cells is used as observations and error assessment is performed by calculating the difference between the reference model—in this case the satellite image—and the one derived from the interpolated observations. Consequently, the error assessment can be used to evaluate the quality of the chosen interpolation method as a whole.

In analogy to the variation of methods and parameters (Section 5.5) carried out for finding the best variogram fitting configuration (Section 7.2.1), a list of options is set up in Table 7.4.

Again, the options the table contains are systematically combined to cover all possible configurations. In addition to the parameters used in Section 7.2.1, the type of the covariance function chosen *cov_fnc* had to be varied in order to allow for better adaptation to the specifics of the given image. On the other hand, the parameter *split_dim* for selecting the split dimension is not necessary here because with a purely spatial variogram there is only *one* dimension the splitting hyperplane can be moved along. Therefore, together with the other options, the number of variants also sums up to 108.

Again, searching for the optimal configuration would be rather cumbersome without the tool introduced in Section 5.5.

Table 7.4: Process method variants for interpolation of sea surface temperature.

Process	Parameter	Variants				Number
aggr	split_pos	mid	med	mea		3
	aggr_pos	mid	med	mea		3
vrgr_fit	cov_fnc	sph	exp	gau		3
	wght_fnc	equ	lin	log	sin	4
					Total:	108

7.7.2 Results

In analogy to the evaluation in Section 7.2.1, the results of the 108 simulations are assessed by plotting the primary quality indicator RMSE against two other indicators: the residuals from the Gauss-Newton fitting procedure (*RMSE_GN*) and the *range* value (*RNG*).

As Figure 7.20 reveals, there is no strong correlation between *range* and RMSE as has been in the experiment of Section 7.2.1. Unlike in that experiment, we do not find such a distinct pattern of *range* for the satellite image. Phenomena like circulation cells and eddies [110, 99] partly disturb the continuity of the system and also the strong correlation between estimated range and interpolation quality.

The 15 variants with lowest RMSE that are listed listed in Table 7.5 reveal the superiority of the exponential covariance function, the logarithm-based weighting function (*wgt_fnc*) and the median value for both partitioning parameters *split position* (*split_pos*) and *aggregation position* (*aggr_pos*).

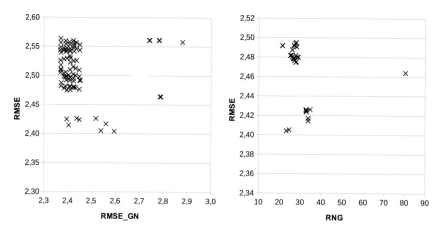

Figure 7.20: Evaluation diagrams of 108 parameter option variants with RMSE values plotted against the residuals from Gauss-Newton optimization *RMSE_GN* (l), and against the range *RNG* (r).

Table 7.5: Listing of the 15 of 108 configuration variants with the lowest RMSE.

nr	split_pos	aggr_pos	wgt_fnc	cov_fnc	rng	rmse	rmse_gn
1	med	med	log	exp	23,33	2,40	2,60
2	mea	mid	log	exp	24,70	2,41	2,54
3	med	mid	sin	exp	33,84	2,41	2,40
4	med	mid	lin	exp	33,80	2,42	2,56
5	mid	mid	log	exp	32,99	2,42	2,45
6	med	mid	log	exp	32,81	2,42	2,40
7	med	mea	log	exp	32,75	2,43	2,44
8	mid	mid	sin	exp	34,76	2,43	2,52
9	mid	mea	equ	exp	80,48	2,46	2,79
10	mid	mea	lin	exp	80,48	2,46	2,79
11	mid	mea	log	exp	80,48	2,46	2,79
12	mid	mea	sin	exp	80,48	2,46	2,79
13	mid	med	equ	exp	80,48	2,46	2,79
14	mid	med	lin	exp	80,48	2,46	2,79
15	mid	med	log	exp	80,48	2,46	2,79

Figure 7.21 illustrates the fitting of the variogram model to the aggregated points from the experimental variogram that yields the smallest RMSE when applied for interpolation (first row of Table 7.5).

The point distribution reveals the striking pattern of a decreasing semivariance for big distances, which indicates the anomaly of non-stationarity caused by oceanic fronts, as already mentioned above. Apart from the discontinuities in the upper half of the satellite image (Figure 7.19), there are large areas of similar

Variogram generated by parameters:
split_dim= max_rel_dev, split_pos= med, aggr_pos= med, wgt_fnc= log, cov_fnc= exp

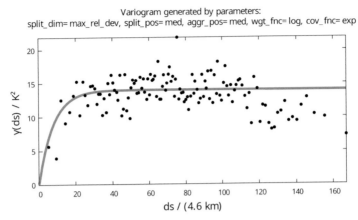

Figure 7.21: Variogram generated by the random observations of the sea surface temperature (SST) image; the distance unit 4.6 *km* results from the pixel-wise treatment of the data, the unit K^2 is due to the square expression within the variogram (Equation 3.2).

value which are distributed all over the region. So there is a considerable amount of rather distant pairs of observations with semivariances which are smaller than the average, which leads to the variogram pattern as found in Figure 7.21.

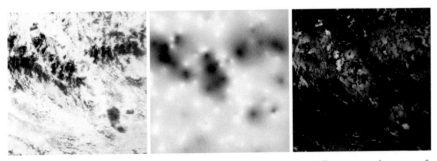

(a) Reference satellite image (see Figure 7.19 for source specification)

(b) Grid derived from interpolation of samples by kriging

(c) Difference map between reference image and interpolated grid

Figure 7.22: SST satellite image on which 200 random observations were carried out for interpolation by kriging; from that, a difference map can be derived.

7.7.3 Conclusions

As the kriged field grid in Figure 7.22(b) shows, the interpolated model does not seem to represent the patterns which can be identified accordingly in the reference satellite image. There are interpolation artefacts as well manifested as *prussian helmets* [129, p. 39], indicating some degree of incompatibility between the observed phenomenon and the interpolated model. This discrepancy also induces the noticeable textured difference image (Figure 7.22(c)).

These negative quality indicators do not, however, disqualify the interpolation method as a whole. It can still provide a valid representation of the phenomenon in terms of minimum RMSE towards the reference, which was a target indicator used here.

When an authentic realization according to the covariance structure underlying the phenomenon is prioritized, a conditional simulation [19, p. 190] might be preferred. If the focus is to estimate the average temperature of a region—e.g. for energy calculations in hurricane models [90]—a target indicator like the deviation of the mean should be considered in search of appropriate methods and parameters.

So the selection of the *appropriate interpolator* depends on the objectives and circumstances of the monitoring as much as on the phenomenon itself [99, p. 160]. When observing a high resolution reference model and comparing the interpolated model with it, as was carried out in this experiment, the interpolation method can adapt to those conditions.

While following the paradigm of the "closed loop" [116, p. 9], the framework presented here allows for variation of reference models, interpolation methods, parameter options, and indicators for quality and efficiency in order to address various objectives.

Chapter 8

Conclusions

CONTENTS

8.1   Subsuming System Overview 170
8.2   Perspective 175

8.1   Subsuming System Overview

The monitoring of continuous phenomena is a complex and challenging task on many levels. In this work, a holistic concept has been proposed to divide this task into small cohesive units of operation. They have to be processed sequentially and are interdependent since each process step uses the output of its predecessor as input. There are unlimited options of configuration within this process chain to control its behaviour.

In order to automate and standardize the process of continuous improvement, a generic model for systematic variation of methods and parameters has been worked out. For evaluation of these generated variants, diverse indicators have been identified and specified, of which model quality and the computational workload are the ones that were examined closer here.

The main contributions of this work to the subject area of monitoring continuous phenomena are summarized below. The toolset for variation and evaluation was used throughout the experiments that were carried out to evaluate the proposed concepts based on key indicators. The framework which is relied on in this work constitutes a laboratory for continuous systematic improvement of environmental monitoring.

Minimum Sampling Density Estimation (Sections 5.3.2, 7.1)

In order to estimate the average sampling density that is sufficient to capture a particular phenomenon, a formula was deduced that derives this density from the extent of the observed area and the dynamism parameter *range*. It presumes the parameter to be known or to be estimated from initial observations.

Variogram Fitting (Sections 5.3.5, 7.2)

Variogram fitting is the key task of geostatistics since it adapts the algorithm to the actual statistical properties of the observational data. A new and generic method based on binary space partitioning (BSP) was proposed to aggregate variogram points of arbitrary dimensionality in order to unburden the subsequent parameter fitting procedure.

Model Merging (Sections 5.4.2, 7.3)

The merging of several grid models of one spatial region addresses two challenges which are common for monitoring systems: (1) the continuous and smooth update of a real-time model by new observations and (2) the handling of large sets of observations by subdivision (divide-and-conquer approach). The method exploits the kriging variance to define reasonable weights for grid cells of the source models by which the output models are created.

Compression of Sensor Data (Sections 5.4.3, 7.4)

An algorithm for compression and progressive retrieval of observational data of arbitrary dimensionality was proposed. Its predominant aim is to improve the sufficiency of sensor data transmission and archiving. In progressive mode, it provides coarse values of low data volume and increases in accuracy with each transmission step.

Quantification of Computational Workload (Section 5.5.2)

Limited computational resources are often a crucial issue, especially for wireless sensor networks, environments with real-time requirements, and large datasets. A strategy for machine-independent description of computational workloads was developed and tested. It keeps track of the number of CPU cycles while differentiating portions of code *capable* and *not capable* of multithreading. While the conventional quantification by execution time ignores the parallelization facilities of both software and hardware, the proposed approach provides much more sophisticated information about the scalability of an implementation.

Systematic Variation and Evaluation of Configurations (Section 5.5.3)

The tools described above entail unlimited potential for variation and configuration in order to adapt to the observed phenomena and to thus provide optimal interpolation results while efficiently using the computational resources. When combining the variations within a process chain, the number of possible configurations to test and evaluate might quickly multiply to large numbers. To handle this complexity within a simulation framework is then a problem in its own right. A generic and coherent architecture to switch between methodological variants or to iterate numerical parameters was designed. Systematic iterative improvement of arbitrarily complex monitoring configurations is thus facilitated.

Higher Level Queries (Chapter 6, Section 7.6)

In many cases, the actual rationale behind monitoring continuous phenomena is to retain or reject a hypothesis like an exceeded threshold for the average emission of an air pollutant. A conceptual framework addressing this problem is introduced and evaluated experimentally. It abstracts from the particular interpolation process by utilising a generic raster-vector-interoperability. The kriging variance is exploited as a confidence indicator for an aggregation statement. The representativeness of this indicator is inspected experimentally.

    The features above have been implemented to plan, perform, provide, archive, evaluate and continuously optimise environmental monitoring processes and their results. Most of the proposed algorithms have been tested and evaluated with the support of the tool for systematic variation and evaluation. Besides using synthetic models as a reference, the algorithms were also applied to real world data, namely a satellite image for sea surface temperature (see Section 7.7).

The concept of quantification of computational workload was tested for the computing-intensive task of random field generation.

In the course of the work on the experimental framework there was a constant evolution towards increasingly abstract concepts to describe and quantify the manifold aspects of environmental monitoring. This is the case for *input* parameters like phenomenon dynamism, sampling density, or possible transmission bandwidth. The *system* might be specified by its algorithms, parameters, workload, computer power, storage space, energy demand, and response time. The generated *output* will predominantly be evaluated by its accuracy and resolution. For a thorough planning and/or evaluation of a monitoring system, a systematic consideration of these issues will certainly be helpful (see Section 5.5.4).

From an even more general viewpoint, all the considerations above boil down to a single question: What degree of knowledge about a particular phenomenon is obtainable by what dedication of resources? As a matter of fact, the concepts and solutions that were introduced in this work tackle this central question.

The proposed solutions and evaluation tools do undoubtedly leave room for further investigation and improvement. The intention of the work was not to thoroughly investigate one narrow problem area, but to aim at several challenges that are specific for the monitoring of continuous phenomena. For systematic evaluation of the efficiency of various solutions, a generic framework is provided. New methods with associated parameter settings can easily be integrated and evaluated by using the present architecture. The circular arrangement of the process chain—with genuine RMSE between synthetic and derived model—allows for iterative investigation and improvement of the monitoring task as a whole.

There are many challenges yet to be overcome on the way towards a comprehensive, generic and consistent architecture for environmental monitoring. In the long-term perspective, one might envision a standardized and interoperable infrastructure that mediates between the available sensor observations and an adequate representation of the phenomenon. Given the insights of this work, the prerequisites for such an environment are listed below.

Interpolation Method Consolidation

Depending on the phenomenon observed, the aims of monitoring and the resources available to achieve it, the method of interpolation has to be chosen carefully. Because of the sheer complexity of the task of interpolation and the plethora of methods available to address it, it is difficult in most cases to decide which one serves the given objectives best. Much specific knowledge and experience is necessary to come to well-founded decisions here.

There are many works dealing with sensor observations and interpolation with different methods and/or parameters. Evaluation is then typically carried out by metrics produced by methods like cross-validation [19, 43, 89]. The problem with these approaches is that cross-validation is not always an indicator for interpolation quality or, as Oliver & Webster point out [95, p. 68]: "The results of cross-validation do not necessarily resolve or justify a choice of model." Continuous random fields and simulation scenarios provide an alternative way of evaluation of methods and parameters.

There is, of course, the disadvantage that the validity of the synthetic models might be doubted. This does not mean, however, that it precludes to draw valid conclusions from such models that are useful in practice. This is the case because there is no absolute model validity anyway [73, p. 247] and real environmental phenomena do only approximately represent random processes [48].

In view of the vast amount of interpolation methods and their variants and associated parameters, there ought to be some mapping policy that relates phenomenon characteristics to interpolation variants which best adapt to those characteristics in particular. A simulation framework with specialized features for systematic variation of both the phenomenon and the interpolation method facilitates the necessary steps towards this goal. The present work might contribute some useful approaches to this endeavour.

Formal Method Specification

As mentioned above, there is an ever-growing amount of interpolation methods and associated variants and parameters, including input and output data formats. The way these specifics are addressed will substantially depend on the particular system and the developers' view on the problem that is factually formulated and implemented by the conventional method.

These circumstances might make it difficult to reproduce a particular processing scenario when a different software product has to be used for some reason. Even more effort might be necessary when the interpolation task is supposed to be run using software as a service (SaaS) where eventually an established interface has to be changed.

One approach to address such problems would be to pursue some degree of interoperability by working out an abstract and standardized definition of the process of interpolation. Even if such a standard would not be directly implemented by software suppliers, it would at least provide a common ground for communication about how a particular implementation works instead of more or less considering it a black box.

Such a common ground specification will of course presume the agreement between its necessity and general structure. The observable trend of cloud computing with its associated advantages might foster developments towards such an agreed specification.

Field Data Type for Data Provision

Observations of continuous phenomena are of specific character depending on the circumstances under which they have to be carried out. More often than not, they do not cover the points or regions which are of interest for the task or question at hand. So instead of the original observational data, applications rather need estimations of the observed variable at arbitrary points or regions in space and time. This is crucial to perform any analysis that is based on the combination of the observed value with some other occurrence, often in order to identify any causative interaction [31, p. 32].

To support such analyses or also for ad hoc queries in space and time, it is necessary to provide an infrastructure that can serve as a mediator between raw data and expected queries. One important means to address this task is the introduction of a specific field data type that is intended to represent continuous phenomena [76, 20, 25].

However, the described concept of a field data type presumes the attributes already mentioned: the consolidation and the standardised description of methodological variants of interpolation.

In an ideal scenario, such a system is continuously fed with observations and simultaneously provides a real-time model or historical data via a standardized interface. Apart from interpolated and cached data to speed up queries, the observational data could be stored without redundant interpolations. Thus, the model of the phenomenon can be provided dynamically at arbitrary points or regions as a function of original observations and the associated interpolation specifications. Such an infrastructure would significantly increase the usability of available observational data and therefore widen the range of their utilization.

Complex Event Definition

Another feature that was addressed in this work is the definition of complex events which are associated with a continuous dynamic field. It entails the specification of a spatio-temporal region for which some condition about the observed value has to be confirmed or rejected. Exceeding a daily maximum value of a particular pollutant in a city district is an example. The definition of such a hazardous situation as a hypothesis that has to be rejected permanently [55, p. 24] would entail insufficient observations as a critical event. So a high regional kriging variance due to sensor failure would trigger an alarm just as an exceeded threshold would.

Furthermore, the spatio-temporal regions under investigation could be more complex than static areas. A trajectory might also be used to interact with the model. For example, the radiation that will be accumulated by a vehicle during a planned mission in a contaminated region could be estimated by spatio-temporally intersecting the trajectory with the model and aggregating the radioactive exposure.

So when given an appropriate model, the continuous phenomenon just means that—in a figurative sense—it is possible to place or move virtual sensors arbitrarily in the region that is sufficiently covered by interpolation. This kind of continuous replication of the observed field is a powerful approach wherever flexible usage of the variable of interest is needed. Other representations like grids or isolines are more limited. However, they can of course easily be derived from the field data type at an arbitrary resolution. So the concept of a virtual sensor as introduced in Chapter 6 might serve as a useful and generic concept in this context.

8.2 Perspective

From the features listed above, each single one is more or less dependent on its predecessors in the given order. Although these features are interdependent in operation, the yet unsolved problems which they contain can—at least to a certain degree—be demarcated through abstraction and thus be worked on individually. This provides plenty of material for challenging scientific work in this area.

The circular principle of simulated monitoring using meaningful quality indicators as presented in this work can easily be extended to develop and evaluate new approaches. The continuous checking against a complex state definition like proposed in Chapter 6 and applied in Section 7.6 can be thought of as integrated in a live monitoring system representing a watchdog functionality. The specifics of the state to be checked (region, time interval, variable, threshold, confidence level) can be defined regardless of the observation arrangement, the interpolation method and the data format.

This scenario is an example of how knowledge about a continuous phenomenon is expressed on a higher level of abstraction. Instead of dealing with individual sensor observations, the context here is aggregations and probabilities or confidence estimations. This decoupling of concerns helps to pursue a modular system architecture and thus facilitates the evolution of involved modules. They become self-contained and interchangeable since they are defined abstractly by their functionality and their interfaces.

From the current developments of environmental monitoring it can be concluded that for responsive systems such an abstraction will become more common. The process of interpolation itself will further diversify according to methods and variants and a main challenge will be to assess the dynamism of a phenomenon by observations and consequently choose the appropriate interpolation method with appropriate associated parameters. Extensive experimental work will help to find and formulate general rules to govern this decision process.

On the input side of monitoring, there has been much research on behalf of network organization and data transmission strategies. The efficiency of this component and the subsequent data processing will remain subject to continuing investigation and improvement.

On the output side, an increasing degree of abstraction with respect to the representation of knowledge about continuous phenomena becomes apparent while regarding concepts like the field data type or aggregations derived from it.

Future developments might integrate data stream management, selection and execution of appropriate interpolation algorithms and parameters, sensor data management with the help of the field data type, provision of a query language suited for the context of continuous phenomena, and, an infrastructure based on that for critical state detection and notification.

To address this problem field in an appropriate and substantial manner, several conflicting requirements need to be carefully balanced: capability, performance, efficiency, interoperability, extensibility, ease of use, credibility and popularity. The priorities of these requirements will change several times during a system's life cycle.

With the increasing availability of environmental observations and the growing demand for actual knowledge derived from them [27], it is just a question of consequential reasoning to come to a compound of modular solutions as suggested in this work. Just like with other subject areas in the realm of geographic information science, the approaches for the monitoring of continuous phenomena will have to undergo a continuous process of consolidation and specification before becoming a ubiquitous standard.

References

[1] Petter Abrahamsen. 1997. *A review of Gaussian random fields and correlation functions*. Norsk Regnesentral/Norwegian Computing Center.

[2] Agterberg, F.P. 1974. *Geomathematics: Mathematical Background and Geoscience Applications (Development in Geomathematics)*. Elsevier Science Ltd.

[3] Mohamed Hossam Ahmed, Octavia Dobre and Rabie Almatarneh. 2012. Analytical evaluation of the performance of proportional fair scheduling in ofdma-based wireless systems, 08 2012.

[4] Martin Aigner and Bert Jüttler. 2009. Robust fitting of implicitly defined surfaces using gauss–newton-type techniques. *The Visual Computer* 25(8): 731–741, Apr. 2009.

[5] Andradóttir, S. 1998. Simulation optimization. pp. 307–334. *In*: Jerry Banks (ed.). *Handbook of Simulation: Principles, Methodology, Advances, Applications, and Practice*. Wiley-Interscience.

[6] Annalisa Appice, Anna Ciampi, Fabio Fumarola and Donato Malerba. 2014. *Data Mining Techniques in Sensor Networks*. Springer London.

[7] Mustafa M. Aral. 2011. *Environmental Modeling and Health Risk Analysis (Acts/Risk)*. Springer-Verlag GmbH.

[8] Margaret Armstrong. 1998. *Basic Linear Geostatistics*. Springer Nature.

[9] Jerry Banks (ed.). 1998. *Handbook of Simulation: Principles, Methodology, Advances, Applications, and Practice*. Wiley-Interscience.

[10] Remi Barillec, Ben Ingram, Dan Cornford and Lehel Csató. 2011. Projected sequential gaussian processes: A c++ tool for interpolation of large datasets Solutions with heterogeneous noise. *Computers & Geosciences* 37(3): 295–309. Geoinformatics for Environmental Surveillance.

[11] Mike J. Barnsley. 2007. *Environmental Modeling: A Practical Introduction*. CRC Press.

[12] Paul Mac Berthouex and Linfield C. Brown. 1994. *Statistics for Environmental Engineers*. CRC Press.

[13] Keith Beven. 2009. *Environmental modelling: An uncertain future?* CRC Press.

[14] Louis G. Birta and Gilbert Arbez. 2007. *Modelling and Simulation: Exploring Dynamic System Behaviour*. Springer.

[15] Jonathan D. Blower, A.L. Gemmell, Guy H. Griffiths, Keith Haines, Adityarajsingh Santokhee and Xiaoyu Yang. 2013. A web map service implementation for the visualization of multidimensional gridded environmental data. *Environmental Modelling & Software* 47: 218–224.

[16] George E.P. Box and Norman R. Draper. 2007. *Response Surfaces, Mixtures, and Ridge Analyses*. John Wiley & Sons Inc.

[17] Thomas Brinkhoff. 2013. *Geodatenbanksysteme in Theorie und Praxis*. Wichmann Herbert.

[18] Robert M. Brunell. 1992. An automatic procedure for fitting variograms by cressie's approximate weighted least squares criterion. *Department of Statistical Science Technical Report No. SMU/DS/TR, Southern Methodist University*.

[19] Peter A. Burrough, Rachael A. McDonnell and Christopher D. Lloyd. 2015. *Principles of Geographical Information Systems*. Oxford University Press.

[20] Gilberto Camara, Max J. Egenhofer, Karine Ferreira, Pedro Andrade, Gilberto Queiroz, Alber Sanchez, Jim Jones and Lubia Vinhas. 2014. Fields as a generic data type for big spatial data. In *International Conference on Geographic Information Science*, pp. 159–172. Springer.

[21] Jean-Paul Chiles and Pierre Delfiner. 2012. *Geostatistics: Modeling Spatial Uncertainty*. John Wiley & Sons, Inc.

[22] Chun, Y. and D.A. Griffith. 2013. *Spatial Statistics and Geostatistics: Theory and Applications for Geographic Information Science and Technology*. SAGE Advances in Geographic Information Science and Technology Series. SAGE Publications.

[23] Thomas H. Cormen, Charles E. Leiserson, Ronald L. Rivest and Clifford Stein. 2005. *Introduction to Algorithms*. The MIT Press.

[24] Dan Cornford, Lehel Csató and Manfred Opper. 2005. Sequential, bayesian geostatistics: a principled method for large data sets. *Geographical Analysis* 37(2): 183–199.

[25] Helen Couclelis. 1992. People manipulate objects (but cultivate fields): beyond the raster-vector debate in gis. *Theories and Methods of Spatio-temporal Reasoning in Geographic Space*, pp. 65–77.

[26] Thomas J. Cova and Michael F. Goodchild. 2002. Extending geographical representation to include fields of spatial objects. *International Journal of Geographical Information Science* 16(6): 509–532.

[27] Max Craglia, Kees de Bie, Davina Jackson, Martino Pesaresi, Gábor Remetey-Fülöpp, Changlin Wang, Alessandro Annoni, Ling Bian, Fred Campbell, Manfred Ehlers, John van Genderen, Michael Goodchild, Huadong Guo, Anthony Lewis, Richard Simpson, Andrew Skidmore and Peter Woodgate. 2012. Digital earth 2020: towards the vision for the next decade. *International Journal of Digital Earth* 5(1): 4–21.

[28] Noel Cressie. 1985. Fitting variogram models by weighted least squares. *Journal of the International Association for Mathematical Geology* 17(5): 563–586, Jul. 1985.

[29] Noel Cressie. 1990. The origins of kriging. *Mathematical Geology* 22(3): 239–252, Apr. 1990.

[30] Noel Cressie. 1993. *Statistics for Spatial Data*. John Wiley & Sons Inc.

[31] Noel Cressie and Christopher K. Wikle. 2011. *Statistics for Spatio-Temporal Data*. John Wiley and Sons Ltd.

[32] Thanh Dang, Nirupama Bulusu and Wu-chi Feng. 2013. Robust data compression for irregular wireless sensor networks using logical mapping. *ISRN Sensor Networks*, 2013.

[33] Peter A.M. de Smet, Jan Horálek and Bruce Denby. 2007. European air quality mapping through interpolation with application to exposure and impact assessment. *In*: Arno Scharl and Klaus Tochtermann (eds.). *The Geospatial Web: How Geobrowsers, Social Software and the Web 2.0 are Shaping the Network Society (Advanced Information and Knowledge Processing)*. Springer.

[34] Desassis, N. and D. Renard. 2013. Automatic variogram modeling by iterative least squares: Univariate and multivariate cases. *Mathematical Geosciences* 45(4): 453–470, May 2013.

[35] Barnali Dixon and Venkatesh Uddameri. 2016. *GIS and Geocomputation for Water Resource Science and Engineering*. John Wiley & Sons.

[36] Marco F. Duarte and Richard G. Baraniuk. 2012. Kronecker compressive sensing. *IEEE Transactions on Image Processing* 21(2): 494–504.

[37] Ehlers, M. 2008. Geoinformatics and digital earth initiatives: a german perspective. *International Journal of Digital Earth* 1(1): 17–30.

[38] Eric Evans. 2003. *Domain-Driven Design: Tackling Complexity in the Heart of Software.* Addison-Wesley Professional.

[39] Domenico Ferrari. 1978. *Computer Systems Performance Evaluation.* Prentice Hall.

[40] Paul J. Fortier and Howard E. Michel. 2003. *Computer Systems Performance Evaluation and Prediction.* Digital Press.

[41] Antony Galton. 2019. Space, time and the representation of geographical reality. In *The Philosophy of GIS*, pp. 75–97. Springer.

[42] João Gama and Mohamed Medhat Gaber (eds.). 2007. *Learning from Data Streams.* Springer-Verlag Berlin Heidelberg7.

[43] João Gama and Rasmus Ulslev Pedersen. 2007. Predictive learning in sensor networks. In *Learning from Data Streams*, pp. 143–164. Springer.

[44] Xavier Gandibleux, Marc Sevaux, Kenneth Sörensen and Vincent T'kindt (eds.). 2004. *Metaheuristics for Multiobjective Optimisation.* Springer Berlin Heidelberg.

[45] Roman Garnett, Michael A. Osborne and Stephen J. Roberts. 2010. Bayesian optimization for sensor set selection. pp. 209–219. *In*: Tarek F. Abdelzaher, Thiemo Voigt and Adam Wolisz (eds.). Proceedings of the 9th International Conference on Information Processing in Sensor Networks, IPSN 2010, April 12–16, 2010, Stockholm, Sweden. ACM.

[46] Andrew Gelman, John B. Carlin, Hal S. Stern, David B. Dunson, Aki Vehtari and Donald B. Rubin. 2014. Bayesian Data Analysis, volume 2. CRC Press Boca Raton, FL.

[47] John P. Van Gigch. 1991. *System Design Modeling and Metamodeling.* Springer US.

[48] Michael E. Ginevan and Douglas E. Splitstone. 2004. *Statistical Tools for Environmental Quality Measurement.* Chapman and Hall/CRC.

[49] Rafael C. Gonzalez and Richard E. Woods. 2002. *Digital Image Processing (2nd Edition).* Prentice Hall.

[50] Michael F. Goodchild. 2010. Twenty years of progress: Giscience in 2010. *Journal of Spatial Information Science* 2010(1): 3–20.

[51] Hugues Goosse. 2015. *Climate System Dynamics and Modelling*. Cambridge University Press.

[52] Benedikt Gräler, Edzer Pebesma and Gerard Heuvelink. 2016. Spatio-temporal interpolation using gstat. *R Journal* 8(1): 204–218.

[53] Benedikt Gräler, Mirjam Rehr, Lydia Gerharz and Edzer Pebesma. 2012. Spatio-temporal analysis and interpolation of pm10 measurements in europe for 2009. *ETC/ACM Technical Paper* 8: 1–29.

[54] Carlos Guestrin, Andreas Krause and Ajit Paul Singh. 2005. Near-optimal sensor placements in gaussian processes. In *Proceedings of the 22nd International Conference on Machine Learning*, pp. 265–272. ACM.

[55] Peter Guttorp. 2001. Environmental statistics. *In*: Adrian E. Raftery, Martin A. Tanner and Martin T. Wells (eds.). *Statistics in the 21st Century*. Chapman and Hall/CRC.

[56] Denis Havlik, Sven Schade, Zoheir A. Sabeur, Paolo Mazzetti, Kym Watson, Arne J. Berre and Jose Lorenzo Mon. 2011. From sensor to observation web with environmental enablers in the future internet. *Sensors* 11(4): 3874–3907.

[57] Katharina Henneböhl, Marius Appel and Edzer Pebesma. 2011. Spatial interpolation in massively parallel computing environments. In *Proc. of the 14th AGILE International Conference on Geographic Information Science (AGILE 2011)*.

[58] Yan Huang, Jingliang Peng, C.-C. Jay Kuo and M. Gopi. 2008. A generic scheme for progressive point cloud coding. *IEEE Transactions on Visualization and Computer Graphics* 14(2): 440–453.

[59] Fulgencio Marín Martínez and Julio Sánchez-Meca. 2010. Weighting by inverse variance or by sample size in random-effects meta-analysis. *Educational and Psychological Measurement* 70(1): 56–73.

[60] Edward H. Isaaks and R. Mohan Srivastava. 1990. *An Introduction to Applied Geostatistics*. Oxford University Press.

[61] Christine Jardak, Janne Riihijärvi, Frank Oldewurtel and Petri Mähönen. 2010. Parallel processing of data from very large-scale wireless sensor networks. In *Proceedings of the 19th ACM International Symposium on High Performance Distributed Computing*, pp. 787–794. ACM.

[62] Edwin T Jaynes. 2003. *Probability Theory: The Logic of Science*. Cambridge University Press.

[63] Guang Jin and Silvia Nittel. 2008. Towards spatial window queries over continuous phenomena in sensor networks. *IEEE Transactions on Parallel and Distributed Systems* 19(4): 559–571.

[64] Jorgensen, S.E. 1994. *Fundamentals of Ecological Modelling, Second Edition (Developments in Environmental Modelling)*. Elsevier Science.

[65] Matthias Katzfuss and Noel Cressie. 2011. Tutorial on fixed rank kriging (frk) of co2 data. *The Ohio State University: Columbus*, OH, USA.

[66] Johnsen Kho, Alex Rogers and Nicholas R. Jennings. 2009. Decentralized control of adaptive sampling in wireless sensor networks. *ACM Transactions on Sensor Networks (TOSN)* 5(3).

[67] Jonathan Gana Kolo, S. Anandan Shanmugam, David Wee Gin Lim, Li-Minn Ang and Kah Phooi Seng. 2012. An adaptive lossless data compression scheme for wireless sensor networks. *Journal of Sensors* 2012.

[68] Werner Kuhn. 2012. Core concepts of spatial information for transdisciplinary research. *International Journal of Geographical Information Science* 26(12): 2267–2276.

[69] Patrice Langlois. 2013. *Simulation of Complex Systems in GIS*. John Wiley & Sons.

[70] Christian Lantuéjoul. 2002. *Geostatistical Simulation*. Springer Berlin Heidelberg.

[71] Craig Larman. 2001. *Applying UML and Patterns: An Introduction to Object-Oriented Analysis and Design and the Unified Process (2nd Edition)*. Prentice Hall PTR.

[72] Stephen Lavenberg (ed.). 1983. *Computer Performance Modeling Handbook (Notes and Reports in Computer Science and Applied Mathematics)*. Academic Press.

[73] Averill M. Law. 2014. *Simulation Modeling and Analysis*. McGraw-Hill Education Europe.

[74] Leinonen, M., M. Codreanu and M. Juntti. 2014. Compressed acquisition and progressive reconstruction of multi-dimensional correlated data in wireless sensor networks. In *2014 IEEE International Conference on Acoustics, Speech and Signal Processing (ICASSP)*, pp. 6449–6453, May 2014.

[75] Jin Li and Andrew D. Heap. 2008. A review of spatial interpolation methods for environmental scientists. Technical report, Geoscience Australia, Australian Government.

[76] Qinghan Liang, Silvia Nittel and Torsten Hahmann. 2016. From data streams to fields: Extending stream data models with field data types. pp. 178–194. *In*: Jennifer A. Miller, David O'Sullivan and Nancy Wiegand (eds.). *Geographic Information Science: 9th International Conference, GIScience 2016, Montreal, QC, Canada, September 27–30, 2016, Proceedings.* Springer International Publishing.

[77] Shan Lin, Fei Miao, Jingbin Zhang, Gang Zhou, Lin Gu, Tian He, John A. Stankovic, Sang Son and George J. Pappas. 2016. Atpc: Adaptive transmission power control for wireless sensor networks. *ACM Transactions on Sensor Networks (TOSN)* 12(1).

[78] Bin Liu, Dawid Zydek, Henry Selvaraj and Laxmi Gewali. 2012. Accelerating high performance computing applications: Using cpus, gpus, hybrid cpu/gpu, and fpgas. In *Parallel and Distributed Computing, Applications and Technologies (PDCAT), 2012 13th International Conference on*, pp. 337–342. IEEE.

[79] Jian Guo Liu and Philippa J. Mason. 2016. *Image Processing and GIS for Remote Sensing: Techniques and Applications.* John Wiley & Sons.

[80] Peter Lorkowski and Thomas Brinkhoff. 2015. Environmental monitoring of continuous phenomena by sensor data streams: A system approach based on kriging. In *Proceedings of EnviroInfo and ICT for Sustainability 2015.* Atlantis Press.

[81] Peter Lorkowski and Thomas Brinkhoff. 2015. Towards real-time processing of massive spatio-temporally distributed sensor data: A sequential strategy based on kriging. In *Lecture Notes in Geoinformation and Cartography*, pp. 145–163. Springer Nature.

[82] Peter Lorkowski and Thomas Brinkhoff. 2016. Compression and progressive retrieval of multi-dimensional sensor data. *ISPRS—International Archives of the Photogrammetry, Remote Sensing and Spatial Information Sciences* XLI-B2: 27–33, June 2016.

[83] Chunsheng Ma. 2007. Stationary random fields in space and time with rational spectral densities. *IEEE Transactions on Information Theory* 53(3): 1019–1029.

[84] Georges Matheron. 1988. *Estimating and Choosing.* Springer.

[85] Steve McKillup and Melinda Darby Dyar. 2010. *Geostatistics Explained: An Introductory Guide for Earth Scientists.* Cambridge University Press.

[86] Henry Ponti Medeiros, Marcos Costa Maciel, Richard Demo Souza and Marcelo Eduardo Pellenz. 2014. Lightweight data compression in

wireless sensor networks using huffman coding. *International Journal of Distributed Sensor Networks* 10(1).

[87] Stephen J. Mellor and Marc J. Balcer. 2002. *Executable UML: A Foundation for Model-Driven Architecture*. Addison-Wesley Professional.

[88] Alfred Mertins. 1999. Signal analysis: Wavelets, filter banks, time-frequency transforms and applications.

[89] Jeffrey C. Meyers. 1997. *Geostatistical Error Management: Quantifying Uncertainty for Environmental Sampling and Mapping (Industrial Engineering)*. John Wiley & Sons Inc.

[90] Michaud, L.M. 2001. Total energy equation method for calculating hurricane intensity. *Meteorology and Atmospheric Physics* 78(1-2): 35–43.

[91] Jennifer A. Miller, David O'Sullivan and Nancy Wiegand. 2016. *Geographic Information Science: 9th International Conference, GIScience 2016, Montreal, QC, Canada, September 27-30, 2016, Proceedings*, volume 9927. Springer.

[92] Werner G. Müller. 1999. Least-squares fitting from the variogram cloud. *Statistics & Probability Letters* 43(1): 93–98, May 1999.

[93] Christian Nagel, Bill Evjen, Jay Glynn, Karli Watson, Morgan Skinner and Allen Jones. 2005. *Professional C# 2005*. Wrox.

[94] Dean S. Oliver. 1995. Moving averages for gaussian simulation in two and three dimensions. *Mathematical Geology* 27(8): 939–960, Nov. 1995.

[95] Margaret A. Oliver and Richard Webster. 2015. *Basic Steps in Geostatistics: the Variogram and Kriging*. Springer-Verlag GmbH.

[96] Michael A. Osborne, Stephen J. Roberts, Alex Rogers and Nicholas R. Jennings. 2012. Real-time information processing of environmental sensor network data using bayesian gaussian processes. *ACM Transactions on Sensor Networks (TOSN)* 9(1): 1.

[97] Michael A. Osborne, Stephen J. Roberts, Alex Rogers, Sarvapali D. Ramchurn and Nicholas R. Jennings. 2008. Towards real-time information processing of sensor network data using computationally efficient multi-output gaussian processes. In *Proceedings of the 7th international conference on Information processing in sensor networks*, pp. 109–120. IEEE Computer Society.

[98] Eric Parent and Etienne Rivot. 2012. *Introduction to Hierarchical Bayesian Modeling for Ecological Data*. Chapman & Hall.

[99] Gongbing Peng, Lance M. Leslie and Yaping Shao (eds.). 2001. *Environmental Modelling and Prediction*. Springer Berlin Heidelberg.

[100] Lluís Pesquer, Ana Cortés and Xavier Pons. 2011. Parallel ordinary kriging interpolation incorporating automatic variogram fitting. Computers & Geosciences 37(4): 464–473.

[101] David Stephen Geoffrey Pollock, Richard C. Green and Truong Nguyen. 1999. *Handbook of Time Series Analysis, Signal Processing, and Dynamics*. Academic Press.

[102] Karl Popper. 2002. *The Logic of Scientific Discovery*. Routledge Chapman Hall. Original Edition: 1959.

[103] Poulton, M.M. (ed.). 2001. *Computational Neural Networks for Geophysical Data Processing (Handbook of Geophysical Exploration: Seismic Exploration)*. Pergamon.

[104] William H. Press, Saul A. Teukolsky, William T. Vetterling and Brian P. Flannery. 2007. *Numerical Recipes 3rd Edition: The Art of Scientific Computing*. Cambridge University Press, New York, NY, USA, 3 edition.

[105] Alan, A. and B. Pritsker. 1998. Principles of simulation modeling. pp. 31–54. *In*: Jerry Banks (ed.). *Handbook of Simulation: Principles, Methodology, Advances, Applications, and Practice*. Wiley-Interscience.

[106] Rasmussen. 2006. *Gaussian Processes for Machine Learning*. MIT University Press Group Ltd.

[107] Philippe Rigaux, Michel Scholl and Agnes Voisard. 2001. *Spatial Databases*. Elsevier Science & Technology.

[108] Christian P. Robert and George Casella. 1999. *Monte Carlo Statistical Methods (Springer Texts in Statistics)*. Springer Verlag.

[109] Pedro Pereira Rodrigues, João Gama and M. Gaber. 2007. Clustering techniques in sensor networks. *Learning from Data Streams*, pp. 125–142.

[110] Floyd F. Sabins. 1996. *Remote Sensing: Principles and Interpretations*. W. H. Freeman.

[111] Hanan Samet. 2006. *Foundations of Multidimensional and Metric Data Structures*. Elsevier LTD, Oxford.

[112] Saket Sathe, Thanasis G. Papaioannou, Hoyoung Jeung and Karl Aberer. 2013. A survey of model-based sensor data acquisition and management.

[113] Klaus Schittkowski. 2002. *Numerical Data Fitting in Dynamical Systems*. Springer.

[114] Connie U. Smith. 2007. Introduction to software performance engineering: Origins and outstanding problems. *In*: Marco Bernardo and Jane Hillston (eds.). *Formal Methods for Performance Evaluation: 7th International School on Formal Methods for the Design of Computer, Communication, and Software Systems, (Lecture Notes in Computer Science)*. Springer.

[115] Petre Stoica and Randolph L. Moses. 2005. *Spectral Analysis of Signals*. Prentice Hall.

[116] Ne-Zheng Sun and Alexander Sun. 2015. *Model Calibration and Parameter Estimation*. Springer-Verlag.

[117] Walter Sun, M. Cetin, W.C. Thacker, T.M. Chin and A.S. Willsky. 2006. Variational approaches on discontinuity localization and field estimation in sea surface temperature and soil moisture. *IEEE Transactions on Geoscience and Remote Sensing* 44(2): 336–350, Feb. 2006.

[118] Clemens Szyperski. 2002. *Component Software: Beyond Object-Oriented Programming (2nd Edition)*. Addison-Wesley Professional.

[119] Taylor, R.N., Nenad Medvidovic and Eric Dashofy. 2009. *Software Architecture*. John Wiley and Sons Ltd.

[120] Matthew J. Tonkin, Jonathan Kennel, William Huber and John M. Lambie. 2016. Multi-event universal kriging (meuk). *Advances in Water Resources* 87: 92–105.

[121] Muhammad Umer, Lars Kulik and Egemen Tanin. 2009. Spatial interpolation in wireless sensor networks: localized algorithms for variogram modeling and kriging. *GeoInformatica* 14(1): 101–134, Feb. 2009.

[122] Adriaan van den Bos. 2007. *Parameter Estimation for Scientists and Engineers*. John Wiley & Sons Inc.

[123] Timo Vihma, Roberta Pirazzini, Ilker Fer, Ian A. Renfrew, Joseph Sedlar, Michael Tjernström, Christof Lüpkes, Tiina Nygard, Dirk Notz, Jerome Weiss et al. 2014. Advances in understanding and parameterization of smallscale physical processes in the marine arctic climate system: a review. *Atmospheric Chemistry and Physics (ACP)* 14(17): 9403–9450.

[124] Hans Wackernagel. 2003. *Multivariate Geostatistics*. Springer-Verlag GmbH.

[125] Hans Wackernagel and Michael Schmitt. 2001. Statistical interpolation models. *In*: Adrian E. Raftery, Martin A. Tanner and Martin T. Wells (eds.). *Statistics in the 21st Century*, Chapter 10. Chapman and Hall/CRC.

[126] Alexander Christoph Walkowski. 2010. *Modellbasierte Optimierung mobiler Geosensornetzwerke für raumzeitvariante Phänomene*. AKA, Akad. Verlag-Ges.

[127] Martin Walter. 2011. *Mathematics for the Environment*. Taylor & Francis Ltd.

[128] Wenping Wang, Helmut Pottmann and Yang Liu. 2006. Fitting b-spline curves to point clouds by curvature-based squared distance minimization. *ACM Transactions on Graphics* 25(2): 214–238, Apr. 2006.

[129] Webster, R. and M.A. Oliver. 2007. *Geostatistics for Environmental Scientists*. Statistics in Practice. Wiley.

[130] Haitao Wei, Yunyan Du, Fuyuan Liang, Chenghu Zhou, Zhang Liu, Jiawei Yi, Kaihui Xu and Di Wu. 2015. A k-d tree-based algorithm to parallelize kriging interpolation of big spatial data. *GIScience & Remote Sensing* 52(1): 40–57.

[131] Whittier, J.C., Silvia Nittel, Mark A. Plummer and Qinghan Liang. 2013. Towards window stream queries over continuous phenomena. In *Proceedings of the 4th ACM SIGSPATIAL International Workshop on GeoStreaming*, IWGS '13, pp. 2–11, New York, NY, USA. ACM.

[132] Bernard P. Zeigler, Herbert Praehofer and Tag Gon Kim. 2000. *Theory of Modeling and Simulation*. Elsevier Science & Technology.

Index

3d 99

A

a priori 41
active sampling 30
adaptive sampling 14
aggregation 4, 10, 19, 27, 29, 52–54, 56–60,
    70, 74, 77, 79–81, 83–87, 105, 108,
    110, 111, 122–127, 136, 137, 139, 140,
    153–161, 165, 171, 175
agriculture 2, 9, 28
air pollution 3, 22, 23, 27, 28
alert system 31, 108
algorithmic optimization 109
anisotropy 15, 19, 39, 40, 42, 44, 47, 50, 69,
    79, 126, 133
approximation 47, 59, 69, 76, 77, 91, 92,
    96, 99, 112, 113, 131
argos (buoys) 145, 146
astronomy 2, 9
atmosphere 117
autocorrelation 4, 14, 18, 19, 38–40, 45, 70,
    75, 80

B

bandwidth 105, 116, 117, 172
beaufort 101
bias 18, 29, 156, 164
binary large object (BLOB) 98, 150
binary space partitioning (BSP) 80–85, 100,
    101, 137, 140, 148, 170
binary tree 102
block kriging 45
bounding box 99, 100

box-muller 70
bsp-tree 81–84, 100

C

calibration 3, 153
central processing unit (CPU) 31, 32,
    109–113, 115, 116, 122, 151, 171
circulation cell 165
classical statistics 125
climatology 9, 163
clouds 77, 79, 80, 84, 87, 137, 138, 162,
    164, 173
cluster computing 108
coding/decoding 100, 104
cokriging 45
communication device 117
communication networks 107
complex state definition 175
composite pattern 114
compression 4, 26, 32, 56, 64–67, 90, 91,
    98–100, 102–105, 107, 116–118, 145,
    148, 150, 151, 171
compression format 103, 104
compression ratio 99, 148, 150
computational complexity 95, 97, 99, 144
computational cost 65, 100, 110, 111
computational efficiency 31, 58, 116
computational effort 5–7, 24, 32, 35, 56, 92,
    97, 108, 120–123, 127, 151
computational workload 25, 31, 46, 48,
    68, 91, 109–111, 120, 122, 151, 152,
    170–172
conditional simulation 17, 168
confidence interval 10, 75, 88

confidence level 144, 175

continuous environmental phenomena 17, 23, 118

continuous fields 2, 3, 9, 12, 17, 19, 48, 52, 63, 69, 70, 74–76, 79, 99, 100, 117, 131, 142, 156

continuous function 71, 84

continuous phenomena 2, 3, 5–9, 11, 19, 21, 22, 26, 27, 31, 50–61, 63, 70, 75, 76, 89, 91, 98, 117, 119, 121, 125, 127, 141, 153, 154, 170–176

continuous vs. discrete 17

correlation 2, 4, 14, 19, 27, 38–42, 45, 46, 50, 59, 71, 72, 78, 99, 100, 126, 127, 135, 137, 139, 159, 161, 165

correlation decay 40, 42

covariance 2, 13–15, 17, 18, 38, 40–48, 50, 71–73, 77, 80, 85, 88, 92, 95, 133–136, 155, 156, 165, 168

covariance function 2, 13–15, 17, 18, 38, 40–44, 47, 50, 71–73, 77, 80, 85, 92, 95, 133–136, 155, 156, 165

covariance matrix 14, 18, 38, 46, 50, 88, 95

covariance model 18

covariate 14

coverage 9, 10, 28, 34, 56, 57, 78, 107, 125, 126, 159, 164

cross-correlation 45

cross-validation 60, 67, 172

cuboid 11

cyclic scale 60, 68

D

data acquisition 31

data compression 56, 98

data format 54, 56, 66, 107, 145, 173, 175

data management 26, 101, 176

data mining 141

data model 55

data processing 60, 98, 175

data stream engine 11, 90–92

data stream management 89, 91, 176

data structure 4, 31, 55, 70, 81–83, 98, 105

data type 9, 11, 19, 26, 29, 98, 100–104, 107, 120, 145, 148, 150, 173–176

database management system (DBMS) 11

decision making/decision maker 6, 33, 118, 127, 159

degrees of freedom 31, 32, 52, 127

delays 26, 92, 117

delegate 11

dependent variable 81

deterministic vs. stochastic 17

deviation map 89, 90, 93, 154

diffusion 17

dimensionality 30, 39, 70, 72, 73, 81, 170, 171

disaster management 28

discrete 2, 6, 9, 17, 19, 22, 23, 26, 52–54, 57, 59, 63, 79, 89, 91, 99, 120, 122, 125, 156, 164

discrete cosine transform 99

discrete wavelet transform 99

discretization 117

disjunctive kriging 45

dispersion 5, 18, 40, 41, 47, 73, 77, 78, 81, 84, 93, 157, 158

dispersion variance 40, 41, 47, 73, 77

divide-and-conquer 170

drifting buoys 30, 49, 145, 146

dual kriging 45

dynamism 3, 4, 6, 10, 15, 16, 19, 22, 30, 52, 56, 60, 63, 74, 75, 92, 96, 117, 135, 142, 155–159, 161, 162, 170, 172, 175

E

economics 9, 27, 28, 34, 47

eddies 165

effectiveness 73, 74, 107

efficiency 2, 3, 6–8, 15, 19, 24, 31, 33, 58, 64, 65, 73, 74, 98, 99, 103, 105–112, 116–118, 148, 150, 153, 168, 172, 175, 176

elevation 52

energy 7, 24, 31–33, 64–68, 98, 106–110, 112, 116–118, 122, 168, 172

energy efficiency 33, 98, 112, 116

energy-efficient 31

environment 6, 14–16, 22–25, 28, 49, 50, 55, 64, 66–68, 91, 96, 98, 101, 105, 106, 110, 111, 114–118, 127, 128, 141, 150, 171, 172

environmental monitoring 10, 15, 22, 26, 31, 33, 50, 115, 118, 120, 170–172, 175

environmental prediction 16

environmental protection 34

environmental science 9

environmental variables 9

erosion 17, 52
error assessment 88, 89, 154, 164
estimation variance 5, 26, 48, 57, 88, 108
evaluation 3, 4, 7, 8, 15, 56, 64, 67, 86, 98,
    105, 111, 113–115, 118, 121, 129, 133,
    138–141, 143, 145, 152, 153, 155, 157,
    158, 165, 166, 170–173
event 17, 26, 54, 174
evolution 172, 175
execution environment 110
experiment 16, 78, 88, 118, 127, 131, 133,
    135, 137, 139–141, 144, 145, 151, 152,
    156–158, 161–165, 168, 170
experimental variogram 4, 38–41, 47, 57,
    70, 76, 77, 79–81, 83–85, 105, 135–138,
    154, 160, 166
explanatory variable 42
exponential decay 95
extension 70, 72, 73, 75, 78, 148

F

factorial kriging 46
field data type 11, 19, 26, 29, 107, 120,
    173–176
field programmable gate array (FPGA) 31,
    112
filtering 18, 45, 49, 50, 73, 92, 121, 162
finance 35
fishery 9, 28
fixed rank kriging 46
fluid dynamics 120
forestry 9, 28
framework 3, 7, 8, 16, 24, 33, 35, 40, 42,
    43, 60, 63, 64, 69, 70, 74, 75, 86, 105,
    111, 114, 115, 120, 121, 141, 145, 153,
    162, 164, 168, 170–173
fringe-effect 73
function calls 114, 124
future information deficit 92

G

gauss-newton 5, 68, 69, 85–87, 136, 139,
    140, 155, 165, 166
gaussian noise 18
gaussian process 14, 50, 92, 156
gaussian process regression 14, 50
gaussian random field 18
gaussian variogram 86, 87
geocoding 52, 54

geographic information system (GIS) 27,
    52, 54, 56
geography 9
geology 2, 9, 15
geometric primitives 55
georeference 54
geoscience 50
geosensor 92
geostatistics 3, 13, 15, 17, 39, 59, 74, 77,
    81, 126, 133, 141, 150, 170
gigacycles 111, 112, 151, 152
granularity 4, 6, 54, 60, 67, 125, 150, 151,
    161
graphics processing units (GPU) 31,
    111–113
grid data 9, 90
gstat 70

H

hardware 3, 6, 24, 31–33, 65–68, 107–118,
    151, 153, 171
heuristics 24, 60, 75
hexagonal 30
hexagonal sampling 30
humidity 23
hydrogeology 9
hydrology 9
hyperplane 81, 82, 84, 87, 137, 138, 165
hypothesis 53, 121, 154, 171, 174

I

independent variable 81
indexing 31, 64, 66, 67, 92, 99, 100, 116,
    118, 150, 151
indicator kriging 46
inference 22
information content 75
information deficit 92
infrastructure 11, 28, 35, 108, 153, 172,
    174, 176
input variable 42, 135, 139
interdependencies 29, 34, 35, 40, 116, 118,
    141, 162
interoperability 8–10, 27, 34, 53, 54, 57–59,
    70, 90, 107, 121, 128, 171, 173, 176
interpolation 2–8, 10–14, 18, 19, 23–26,
    29–31, 35, 37–39, 42, 45–48, 50, 52,
    54–60, 63, 64, 67, 70, 71, 74, 78–80,
    88–91, 97, 105, 107, 111, 117, 118,

120–127, 131, 133, 135, 136, 138–140, 142, 144, 153–157, 160–168, 171–176
interval 10, 25, 54, 60, 68, 75, 77, 81, 84, 87, 88, 92, 96, 100–102, 114, 124, 125, 144, 175
interval scale 101
intrinsic stationarity 13, 38
inverse distance weighting 12, 50, 92
irregularly scattered 24
isoline 175
isotropy 74
iterative fitting 41

K

k-d tree 69, 81–83
kelvin 101, 163
kernel 18, 45, 71, 108
kernel function 18
knowledge 5, 6, 15, 16, 22, 23, 25–29, 35, 49, 52, 53, 57–60, 64, 66, 67, 74, 79, 107, 120, 122, 124, 125, 127, 153, 162, 164, 172, 175, 176
kriging 3–5, 7, 8, 13–15, 18, 19, 23, 25, 26, 37, 38, 42, 45–50, 56, 57, 69, 70, 75, 78, 80, 85, 87, 88, 91–93, 95–97, 108, 121, 122, 124, 127, 131, 132, 141, 142, 144, 154–157, 159–162, 164, 167, 170, 171, 174
kriging error 49, 50
kriging error maps 49, 50
kriging variance 5, 14, 25, 48, 49, 50, 56, 88, 91–93, 95, 97, 124, 127, 141, 144, 154–157, 159–162, 170, 171, 174
kriging variants 14, 50
kurtosis 71, 125

L

lagrange multiplier 46
level of detail 3
limited resources 28, 64, 66, 67, 105, 106, 108
linear regression 14, 135
logistic function 86
lossy compression 26

M

machine learning 14, 50
machine-dependent 111

machine-independent 110, 111, 113, 115, 151, 171
management 11, 26, 28, 64, 67, 89, 91, 101, 113, 150, 151, 176
map of the second kind 48, 50, 90, 159
mapping 24, 52, 90, 91, 99, 104, 173
matrix 14, 18, 38, 46, 50, 72, 88, 92, 95, 99
matrix algebra 92, 99
matrix inversion 14
maxima 10
mean 2, 6, 12–14, 17, 26, 27, 29, 35, 46, 50, 56, 58, 59, 63, 66, 67, 69–73, 80, 82–86, 89, 96, 107, 109, 121, 124–127, 136, 140–142, 146, 150, 151, 155–159, 161, 168, 173, 174
mean absolute deviation 125
mean kriging 127, 156, 157, 159, 161
mean squared error 67, 126
measure of confidence 127
measurement scales 58
median 27, 80, 82–84, 86, 125, 140, 165
medicine 2, 9
memory 92, 98, 109, 113, 123
metadata 25, 27, 50, 148, 150, 151
meteorological station 30
meteorology 9
mining 2, 9, 93, 141
mirror image 40, 42, 85
mobile computing 98, 99
mobility 98
mode 26, 27, 33, 35, 67, 68, 118, 125, 150, 151, 171
model accuracy 117, 127, 157, 161
model evaluation 118
model parameters 150
model quality 6, 22, 89, 170
model validity 173
modeling 3, 15, 16, 22, 23, 27, 52, 64, 70, 153, 161
monitoring 2, 3, 6–8, 10, 15–17, 19, 21–35, 48–50, 52–57, 59–61, 63, 64, 66–70, 74, 75, 88–92, 95, 97, 100, 105–108, 110, 112–115, 117, 118, 120–124, 126, 127, 135, 138, 141, 144, 145, 148, 151–154, 156, 159–162, 164, 168, 170–172, 175, 176
movie 11, 27, 70
moving average 4, 18, 71–73, 136, 151
moving average filter 4, 18, 71, 72, 136, 151

multithreaded 152
multithreading 31, 109, 111, 113, 151, 171
multivariate 45

N

nanojoule 112
nanoseconds 103
nested survey 75
network 8, 22, 24, 30, 32, 33, 35, 49, 64,
    74, 107, 110, 112, 113, 128, 150, 153,
    171, 175
nitrogen 75
noise 9, 14, 15, 18, 42, 47, 69–73, 80, 92,
    126, 136, 155, 163
nominal scale 68, 101, 103, 104
non-linear 17, 85, 87
non-parallelizable 110, 111
nonseparable 43, 44, 47
normal distribution 45, 70
nugget effect 15, 40, 42, 47, 126
nugget variance 42, 50
Nyquist-Shannon 75, 76, 133

O

object 52, 54, 81, 98, 112, 114, 150
ocean 163, 164
oceanic front 164, 166
oceanography 164
oceanology 2
octree 99
octree-cells 99
optimization 5, 6, 22, 31–33, 47, 67, 68, 70,
    85, 87, 105, 108, 109, 113, 123, 141,
    153, 160, 166
ordinal scale 101
ordinary kriging 46
output variable 46
oversampling 135

P

parallelizable 32, 110, 111
parallelization 31–33, 111, 112, 116, 118,
    151, 152, 171
parameter estimation 5, 55, 140, 154, 162
parameter fitting 170
parameterization 153
particulate matter 159

patterns 60, 69, 77, 88, 89, 163, 164, 167
performance evaluation 111, 152
performance indicator 3, 6, 35, 64, 68,
    105–107, 113, 115, 121, 153
periodicity 50, 76, 77, 92
permutation 118
petroleum engineering 9
physical 12, 13, 15, 30, 44, 68, 74, 110, 112,
    126, 128, 151
physical process 13, 30, 44, 74
physical resources 110, 128
pipelining 113
point interpolation 48
pointcloud 40, 77, 80, 84, 137, 138
policy 123, 173
political 28, 34
politics 28
pollution control 9
polymorphic function call 114
polynomial 12, 45
population 40
post-processing 55
pre-processing 9
precipitation 10, 26, 45
precision 29, 39, 53, 56, 115
precision farming 53
principal component analysis 46
principal component kriging 46
probability distribution 71
process 2–8, 10, 11, 13–19, 22–24, 26–30,
    32, 33, 35, 38, 40, 42, 46, 50, 52, 53, 55,
    56, 58–60, 64–70, 73, 74, 79, 82, 84–86,
    88, 89, 91–93, 105, 108, 109, 111–113,
    115, 116, 118, 120–123, 127, 135, 136,
    138, 141, 151, 153–156, 159, 160, 165,
    170–173, 175, 176
process model 3, 64
processing unit 31, 35, 64, 65, 108,
    110–112, 117, 128
processor 32, 109–111, 116, 151
product-sum model 44
prussian helmets 167
pseudo-random 70
public health 9

Q

quantile 5, 87, 125
quantitative 7, 98

R

radiation 3, 24, 121, 174
radioactivity 28
rain gauge 52, 142
rainfall 22, 24, 27
random field 3, 4, 13, 16–19, 24, 40, 42, 43,
    45–47, 64, 69–73, 79, 88, 89, 118, 131,
    133–138, 151, 154, 155, 164, 172, 173
random function 15
random noise 14, 126
random process 10, 13, 17, 19, 173
random sampling 131, 137
range 2, 4, 13, 24, 40, 42, 45, 47, 50, 58,
    59, 69, 72, 75–80, 95, 98, 101, 103,
    126, 132–136, 138–140, 146, 150, 151,
    155–158, 163–166, 170, 174
rank 46
raster data 8, 26, 51, 54–59, 65, 66, 123
raster-vector 57, 121, 171
ratio scale 101
redundancy 26, 47, 49, 85, 117
reference model 6, 7, 24, 56, 64, 70, 79, 89,
    126, 127, 136, 140, 142, 143, 154–156,
    158, 159, 161, 163, 164, 168
regional variogram 79
regionalized variable 14, 126
regions of interest 52, 57, 123
regression 12, 14, 46, 50, 135
regression kriging 46
reliability 5, 19, 56, 124, 149, 159
remote sensing 9, 10, 52, 54, 56, 89, 162,
    164
representativeness 53, 75, 127, 156, 171
requirements 2, 24, 25, 28–31, 33, 35, 50,
    52, 56, 57, 66, 67, 74, 89, 92, 95, 99,
    100, 106, 109, 117, 120, 123, 144, 148,
    151, 171, 176
resampling 9, 123
resource requirements 56, 106, 109
resources 6–8, 22, 23, 27–29, 31, 32, 56,
    64–67, 105–110, 117, 118, 120, 122,
    123, 128, 144, 151, 161, 171, 172
response time 34, 100, 110, 117, 118, 172
responsiveness 26, 56, 105, 121, 123, 144,
    151
risk 87, 91
robust estimation 69, 87
root mean squared error 67, 126
root-mean-square 142, 159

S

safety 27
sampling rate 74, 117
scalability 171
scale 42, 46, 58, 60, 68, 70, 92, 101–104,
    137, 152
sea surface temperature (SST) 145, 163,
    165, 167, 171
second-order stationarity 14
security 24, 34
semivariance 15, 39, 41, 47, 77, 80, 81,
    137, 138, 166, 167
sensor data streams 24, 90, 98, 101, 103
sensor networks 8, 22, 24, 30, 32, 33, 49,
    64, 107, 110, 112, 113, 171
sensor web 90, 101, 112
sensor web enablement (SWE) 90
sensors 2–5, 8, 9, 19, 22–26, 28–33, 35, 49,
    52, 55, 60, 64, 74, 76, 90, 92, 98–101,
    103, 104, 107, 108, 110, 112, 113, 116,
    117, 120–124, 127, 128, 150, 153, 159,
    161, 162, 164, 171, 172, 174–176
separable covariance function 41, 43, 136,
    155, 156
sequential gaussian simulation 18
serial algorithm 111
signal processing 4, 18, 75, 99, 133
signal-to-noise ratio 163
sill 40–43, 47, 58, 59, 69, 73, 75, 77, 79,
    126
simple kriging 14, 38, 46
simulated annealing 18
simulation 3, 6, 15–18, 22–24, 27, 31, 33,
    40, 42, 50, 63–70, 75, 105, 106, 109,
    114, 115, 118, 121, 126, 127, 136, 141,
    142, 145, 152, 153, 165, 168, 171, 173
simulation scenario 68, 69, 105, 126, 142,
    145, 173
simulation techniques 18
singlethreaded 151
skewness 69, 71, 84, 125, 150
sliding window 11, 48
social geography 9
software engineering 124
soil moisture 164
soil science 9
space-time-cube 120
spatial autocorrelation 40
spatial correlation 59

spatial interpolation 111
spatial statistics 13
spatial variability 15
spatio-temporal anisotropy 44
spatio-temporal covariance function 15, 41
spatio-temporal indexing 67
spatio-temporal interpolation 11, 35, 37, 107
spatio-temporal kriging 42
spatio-temporal variogram 43, 44, 47, 80
splines 12
splitting hyperplane 82, 165
standard deviation 70, 78, 125, 150, 157, 163
standard error of the mean 125
standardisation 11
state 8, 24, 31, 32, 40, 48, 49, 53, 54, 60, 70, 74, 75, 90–92, 120, 141, 145, 154, 162, 175, 176
state model 48, 141
static vs. dynamic 17
stationarity 13, 14, 18, 23, 38, 70, 71, 166
stationary random function 15
statistical methods 5, 121
statistical moments 13
statistical parameters 46, 50, 82, 126
steepest ascent 141
stochastic process 13
strong stationarity 13
subdivision 49, 87, 98–101, 144, 170
summarization 127
system model 110
system under investigation 16

T

temperature 2, 3, 10, 22, 23, 52, 98, 101, 120, 125, 145, 150, 162, 163, 165, 167, 168, 171
temporal 2–4, 10, 11, 14–17, 22, 25, 27, 32, 35, 37–44, 47, 48, 52, 54, 55, 58, 60, 64, 66, 67, 70, 72, 75, 80, 81, 85, 92, 93, 95, 96, 98–100, 103, 107, 109, 111, 114, 117, 120–122, 124–126, 133, 136–139, 142, 145, 150, 152, 155, 156, 160, 161, 174
temporal autocorrelation 14, 38, 80
theoretical variogram 5, 38, 40–42, 47, 71, 77, 79–81, 85, 87, 137, 159, 160
thiessen polygons 12

threads 110–112, 151, 152
throughput 50, 117
time 2–5, 7–11, 14, 16, 17, 19, 22, 24–27, 29, 31–35, 39–41, 43–45, 47, 48, 50, 52, 54, 57, 60, 63–68, 70, 71, 74, 78, 89–92, 95, 96, 98–101, 103, 105–118, 120, 122, 123, 125, 127, 137, 142–144, 150–152, 154, 155, 161, 170–172, 174–176
time series 26, 39, 54, 99, 100, 137
topography 52, 117
total variance 77, 78
tracking 67
trajectory/trajectories 54, 55, 57, 121–124, 174
transect 39, 54, 55, 57, 121, 122, 125
transformation 58, 99
transition 127
transmission bandwidth 172
transport 56
trend cluster 92
trend surface 12, 15
turning bands 18

U

uncertainty 14, 19, 48, 97, 134, 142, 159
unconditional simulation 17
universal kriging 15, 46

V

validation 22, 23, 60, 67, 89, 172, 173
variability 2, 15, 47, 52, 71
variance 4, 5, 12–14, 18, 25–27, 40–43, 47–50, 56–59, 66, 70, 71, 73, 76–78, 80, 82, 84–86, 88, 90–95, 97, 108, 121, 122, 124–127, 141, 144, 154–157, 159–162, 164, 170, 171, 174
variogram 4, 5, 13, 27, 30, 38–47, 50, 57, 59, 60, 66–68, 70, 71, 73–77, 79–89, 105, 126, 127, 132, 133, 135–141, 150, 151, 154, 155, 159, 160, 162, 164–167, 170
variogram estimation 76, 79, 138, 141, 159
variogram fitting 5, 30, 40, 45, 70, 73–75, 79, 81, 84, 85, 87, 132, 135, 136, 139, 140, 165, 170
vector data 8, 26, 54, 55, 57–60, 66, 90, 117, 123
vector-raster 57, 58

vector-raster-interoperability 57, 58
visualization 10, 26, 52, 99, 104, 105, 124,
    125
voronoi 12

W

weak stationarity 13
weather 10, 27, 28
web 24, 90, 91, 99, 101, 108, 112
web map service (WMS) 90
web mapping 24, 90, 91, 99
web services 90, 91, 108

weighted average 18
white noise 69–73, 80, 136, 155
whole-part-relationships 114
wind speed 10
wireless sensor networks (WSN) 8, 22, 24,
    33, 64, 110, 112, 113, 171
workload (computational) 25, 31, 46, 48,
    68, 91, 109–111, 120, 122, 151, 152,
    170–172

Z

zip (compression) 148